STRUCTURAL MASONRY
AN EXPERIMENTAL/NUMERICAL BASIS FOR PRACTICAL DESIGN RULES

Structural Masonry

An Experimental/Numerical Basis for Practical Design Rules

Edited by

J.G. ROTS

TNO Building and Construction Research, Rijswijk, Netherlands

CRC Press

Taylor & Francis Group

Boca Raton London New York

CRC Press is an imprint of the
Taylor & Francis Group, an **informa** business

Contents

V

Preface

Nowadays structural masonry forms an attractive alternative to modern supporting structures. The material has an aesthetic appearance, is durable and easy to stack at the building site. The units, blocks or elements are available in many types, sizes and colours. Supporting walls, diaphragm walls, pre-stressed cores and pier-wall systems are examples of structural facilities with which structural masonry competes.

Application of structural masonry, however, is hindered by a lack of knowledge of the behaviour of materials and structures. Most of the present-day design and calculation rules are empirical and traditional by nature. These gaps in knowledge were recorded in 1987 by the CUR PC55 pre-advisory committee. Subsequently in 1989 the CUR PA33 pre-advisory committee outlined a new scientific approach to structural masonry with the integration of numerical mechanics and experimental techniques. This resulted in the research programme 'Structural Masonry I' in the period of 1989-1993.

The research was started at the initiative of the Royal Dutch Association of Brick Manufacturers (*Koninklijk Verbond van Nederlandse Baksteenfabrikanten KNB*), as part of the BRIK programme (Unit, Research, Innovation and transfer of Knowledge). From 1990 the research has also been supported by the calcium silicate industry (*Kalkzandsteenindustrie CVK/RCK*). The research was carried out by TNO Building and Construction Research, Eindhoven University of Technology (*TU Eindhoven*) and Delft University of Technology (*TU Delft*). The objective was to create a basis for the development of practical calculation rules to promote the application of structural masonry in the Dutch building practice. The aim was to integrate experimental, numerical and analytical methods, which has been reflected in a committee structure consisting of a steering committee with three sub-committees. The composition of these committees was as follows:

C76 'Structural Masonry' Steering Committee
C.J.M. Schiebroek, Chairman
J. Jongbloed, Secretary
W.J. Beranek
P. de Jong
J.C. Keller
G.M.A. Kusters
M.H.M. Nieuwenhuys
P.D. Rademaker

A.J.M. Siemes
W.R. de Sitter
J.C. Walraven
P. Kole, Coordinator
J.W. Kamerling, Mentor

A33 'Numerical Masonry Mechanics' Committee
W.R. de Sitter, Chairman
J.G. Rots, Secretary
W.G.J. Berkers
A. Bruijntjes
H. Brummel
H.A.J.G. van den Heuvel (from 1992)
G.J. Hobbelman
H.J.M. Janssen
P.J.G. Merks
K.J. Tonkens (until 1992)
P. Kole, Coordinator
J. Blaauwendraad, Mentor

B50, 'Experimental Masonry Research' Committee
A.J.M. Siemens, Chairman
A.P. van der Marel
A. Brink
H. Brummel
J.J.I. Buisman
G.W.J. van Drie
H.A.J.G. van den Heuvel (from 1992)
H.J.M. Janssen
K.J. Tonkens (until 1992)
R. van der Pluijm
A.Th. Vermeltfoort
P. Kole, Coordinator
J.W. Kamerling, Mentor

Besides, there was a CUR C77 'Analytical Masonry Mechanics' Committee in action under the chairmanship of W.J. Beranek.

This report gives a survey of the experimental/numerical build-up of knowledge. The report has been written by J.G. Rots (Chapters 1, 3, 4, 6-8, preface and final editing), R. van der Pluijm and A.Th. Vermeltfoort (Chapter 2) and H.J.M. Janssen (Chapter 5). P.B. Lourenco has contributed to Section 4.2.

The CUR would like to thank the Royal Dutch Association of Brick Manufacturers, the Cooperative Sales Office Calcium Silicate Industry and the Ministry of Economic Affairs (*Ministerie van Economische Zaken*) for their financial support which enabled this research to be carried out.

April 1994
The CUR Board

Summary

Putting units on top of each other, either with or without cohesion via mortar, is a simple though adequate technique that has been successful ever since remote ages. The resulting stack of masonry is well known for reasons of solidity, durability and aesthetics. However, innovative applications of structural masonry are hindered by the fact that the development of design rules has not kept pace with the developments for concrete and steel. The underlying reason is the lack of insight and models for the complex behaviour of units, joints and masonry composite. This has led to the CUR research project 'Structural Masonry I' initiated by the Netherlands unit and calcium-silicate industries. This report gives an overview of the results.

The main objective was to create a basis for a general approach towards structural masonry. This objective has been met via an integration of experimental and computational techniques. Accurate displacement-controlled materials experiments have produced an extensive database of strength, stiffness and softening properties for tension, compression and shear. This data has been transferred into numerical models for simulating the deformational behaviour and failure behaviour of masonry structures. The models have been implemented into finite and distinct element codes and have been subsequently verified against shear wall experiments and analytical solutions for masonry parts.

The utility of research lies in rationalising the engineering design of structural masonry. Existing calculation methods are mainly of an empirical and traditional nature. Starting from the experimental/numerical basis, existing methods are being validated, extended and improved. This process has been started recently and two examples are included in this report. The examples involve the cracking behaviour and movement joint spacing in walls under restrained shrinkage, and the failure behaviour of pier-wall connections. Other applications have been reported separately, amongst which the validation of calculation methods for diaphragm walls, the assessment of the relation between bending strength and uniaxial bond strength, and the critical review of existing test methods.

The main conclusion is that the investment in modelling returns in results for practice oriented cases. The numerical simulations, supported by experimental data, provide new insight into structural behaviour and support the derivation of rational design rules. Recently discovered properties like softening and dilatancy, being virtually absent in the masonry literature so far, play a crucial role in that process.

XI

Symbols

c	Cohesion, bond shear strength
c_u	Cohesion, ultimate value bond shear strength
D	Stress-strain matrix
E	Elastic modulus
f_c'	Compressive strength
f_{ck}'	Characteristic compressive strength
f_{fl}	Bending strength
f_s	Bond shear strength
f_t	Tensile strength (general)
f_t^u	Ultimate value of tensile strength
f_{tb}	Bond tensile strength
f_{tu}	Unit tensile strength
G	Shear modulus
G_f^I	Mode-I fracture energy for tensile fracture
G_f^{II}	Mode-II fracture energy for shear fracture
H	Wall height
K	Stress displacement matrix
k_n	Normal stiffness of interface
k_t	Shear stiffness of interface
L	(crack-free) wall length
R_a	Axial degree of restraint
R_i	Flexural degree of restraint
u	Relative displacement at normal to interface
v	Relative shear displacement at interface
w	Crack width
ε	Normal strain
ϕ	Angle of internal friction
γ	Shear strain
μ	Friction coefficient
ν	Poisson's ratio
ψ	Angle of dilatancy
σ	Normal stress
τ	Shear stress
τ_u	Shear strength
τ_{wr}	Friction at shear stress

When a specific component is referred to (unit, mortar, masonry, crack, and the like) sometimes additional subscripts are used in conjunction with the symbols.

CHAPTER 1

Introduction

1.1 GENERAL

Building in unit bond has been known since ancient times. Innumerable variations have occurred throughout the centuries, influenced by the local supply of raw materials and the country's culture. Strong points in favour of masonry are undoubtedly the aesthetic appearance, the durability and the simplicity of the stacking technique. Supported by slogans like 'Unit is Beautiful', for clay units the aesthetic motives have gradually gained the upper hand over the structural aspects. In contrast to earlier days the load bearing structure of modern buildings is provided by concrete or steel, while clay units are mainly used as facade cladding. Using calcium silicate products is a different matter. This material has primarily a load bearing structural function and has become popular during the last decade owing to the ample supply of blocks and elements combined with mechanical handling at the building site.

The decrease in the market share of clay unit in the load bearing sector can to a considerable extent be attributed to the fact that the technical-scientific development of design rules has lagged behind compared to concrete and steel. This lack of a good basis for design and calculation hinders innovative applications and causes insufficient understanding of damage symptoms. Also in the case of calcium silicate unit many structural questions have remained unanswered and the traditional rules of thumb are inadequate as soon as new applications outside the usual field of experience are discovered. The gravity of the situation has been shown, among other things, by the TNO report 'Safety of masonry constructions – Probabilistic approach' [1]. It showed that traditional calculation methods do not fit in with the actual behaviour of masonry structures. Some examples are the problems concerning bending strength of facades and cavity walls, crack spacing and movement joint spacing under restrained shrinkage or thermal cooling and the failure behaviour of new types of structures such as diaphragm walls in clay unit or glued pier-wall connections in calcium silicate unit.

The above gaps in knowledge have been recorded systematically by the CUR PC55 'Structural Masonry' pre-advisory committee [2]. This was the basis for the CUR PAC4 programme advisory committee of the same name on which a final advice concerning a research programme was drawn up [3]. In that advice this committee states (quotation): '... knowledge of masonry might be present in a few subsectors but due to the unstructured form of description it cannot or can hardly be used to dimension and calculate structures. There is no unity in the principles of mechanics, no

1

unity in the test methods and no unity in the approach to the problem. ...'. Traditional research in the field of masonry mainly took place (and still takes place) through ad hoc tests of structures, so that the result is only valid for that specific structure, with that specific type of unit and joint, those specific boundary conditions and those specific loads. Extrapolation of the results to other situations is not or only to a very limited extent possible. A more general approach is needed. The CUR PA33 'Computational masonry mechanics' pre-advisory committee [4] has in this respect pointed out the new possibilities of numerical simulation techniques. Especially the non-linear finite element method seems to be appropriate, with the salient detail of the natural presence of 'finite elements' in masonry.

Another aspect that is of importance for the future of structural masonry is the increasing internationalisation of the building process. The originally mainly local character of building in units has resulted in a diversity of approaches. National regulations differ in volume and quality to a large extent. In this field the Netherlands does not keep up with countries such as Germany and Great Britain. The recently formulated Eurocode no 6, 'Common unified rules for masonry structures'[5], is a compromise in which the achievements of the countries with a clear tradition in masonry have been integrated. Defence of the typically Dutch building methods in this European circuit is of major importance. This applies, for example, to our slender cavity walls which differ to a large extent from the building practice with solid walls in surrounding countries. This stimulates the Netherlands to increase its research efforts.

The above tendencies have been recognised by the Royal Dutch Association of Unit Manufacturers (KNB) and have led to the CUR 'Structural Masonry I' research project during 1989-1993. As from 1990 the project has also been supported by the calcium silicate (sandlime unit) industry (CVK/RCK).

1.2 PURPOSE AND FRAMEWORK OF THE RESEARCH

Starting point in the research was the above final advice from the CUR PAC4 committee [3]. Following this advice the main purpose of the research can be defined as follows:

To create a bases for a general approach to structural masonry. This includes closing the gap between theory, mechanics models and materials testing on the one hand, and the structural practice on the other hand, which should result in a clearer insight and a scientific basis for the behaviour of masonry structures.

As far as the research method is concerned a combined experimental/numerical approach has been chosen. This method has been recommended by the CUR PA33 pre-advisory committee [4] and is considered to be the most promising to achieve a general approach to structural masonry. The outlines of the method are:

a) Execution of material experiments. Through systematic testing a fundamental knowledge will be developed concerning tensile strength, shear strength, compressive strength and the deformational capacity of masonry as a synthesis of the basic components unit and mortar.

b) Formulation of material models. The acquired material properties are transformed into constitutive models for unit, joint, interface and composite. The models have to correctly represent the non-linear behaviour of the material through crack formation.

c) Implementation of the models in finite element method soft ware. In this project the multi-purpose program DIANA has been chosen because of the ample possibilities in the non-linear field and because of the good access to TNO Building and Construction Research and both Universities of Technology. Besides, partial studies are carried out with the special purpose program UDEC.

d) Verification by means of construction experiments. A limited number of experiments on parts of structures, piers for example, are necessary to verify whether the numerical models are sufficiently reliable. In case numerical results and experimental results do not correspond sufficiently, an adjustment of the models or parameters should take place.

e) Execution of practice-oriented case studies. With the help of new material data and numerical models the behaviour of masonry structures can be simulated. The purpose of which is a clearer insight. The ultimate purpose is the development, foundation and improvement of calculation rules, auxiliary tables and diagrams, as well as the propagation of the obtained knowledge to the building practice.

This report contains the results of the above process. Chapter 2 shows an outline of the materials testing carried out for tension, compression and shear, whereby not only strength criteria are dealt with but also the softening behaviour after having reached the ultimate strength. This softening behaviour is essential with unit-like materials and determines the way in which crack formation propagates within a structure. Subsequently Chapter 3 describes the way in which experimental observations are transformed into cracking and friction models within DIANA. A distinction is made between micro and macro. Micro-models model units and joints separately, whereas macro-models describe the overall behaviour of masonry-as-composite. Given the fundamental purpose of the research the accent has been laying on the micro-models. Chapter 4 deals with evaluation and verification studies for masonry components that are in tension and piers under shear. The numerical results are discussed and judged on the basis of expectations derived from manual calculations and experimental macro-data. Chapter 5 describes the numerical approach with UDEC. With this program also pier studies have been carried out. The distinct element method within UDEC is compared with the finite element method within DIANA.

Subsequently the step towards applications is made in Chapters 6 and 7. Two practice-oriented case studies are discussed: the failure behaviour of pier-wall connections and the crack formation of walls under restrained shrinkage. The case studies illustrate the way in which practical calculation rules for respectively stability and dilatation's (movement joints) are given a basis or are improved via the research carried out. Other applications of numerical simulations have been published separately, among which the formulation of design models for masonry diaphragm walls [6], the assessment of existing test methods for cohesion (tensile and bending) [7] and the evaluation of existing test methods as well as the development of a new test method for shear [8]. Finally Chapter 8 deals with the conclusions.

CHAPTER 2

Testing of materials

2.1 GENERAL

This chapter describes tests by which various material parameters have been determined. Within this research distinction has been made between the components masonry consists of. This level of research is often indicated with micro- or detail level. The tests concerned half-unit masonry with mortar joints.

The main purpose of the tests is to determine parameters which are necessary in finite element method programs to describe the behaviour of materials.

A second purpose of the tests is to determine possible links between the various parameters so that in the future relatively simple tests will be sufficient to determine normative parameters.

A relatively soft mud clay unit, a wire cut unit and a calcium silicate unit, all of standard quality were used. With the compression tests another stronger wire cut unit was used as well. These units were used in combination with calcareous and cement-rich mortars. As a result a wide range of possible combinations of materials was tested, providing insight into the magnitude and possible variation of the material parameters which were determined by means of the various tests.

Deformation-controlled compression, tension and shear tests were carried out. With this way of testing the load is applied in such a way that during a test, deformation of a specimen increases with a prescribed speed over a chosen measuring length; this in contrast to load-controlled tests whereby the load increases with a prescribed speed. The great advantage of such deformation-controlled tests is the possibility to continue registration of the behaviour after the maximum force has occurred.

The next section deals with the materials that were used and the way in which they were processed. Besides discussing the terminology and coding that were used, the following sections subsequently discuss the compression, tension and shear tests.

2.2 MATERIALS, THEIR PROCESSING AND CODING

2.2.1 *Materials*

In all, 4 types of units have been tested, namely moulded red clay units, manufactured by the Vijf Eiken BV, yellow and blue extrusion wire cut units, manufactured by Joosten BV and calcium silicate units. For each test arrangement every type of

4

Table 2.1a. Properties of units used.

Units	Compressive strength (N/mm^2)	Unit mass (kg/m^3)	Dimensions (mm^3)
Joosten (JB)			
(blue extrusion wire cut units)	120	2053	$210 \times 97 \times 50$
Joosten (JG)			
(yellow extrusion wire cut units)	66	1994	$204 \times 98 \times 50$
Vijf Eiken (VE)			
(moulded red clay units)	33	1880	$208 \times 98 \times 50$
Calcium silicate units (CS)	35	1810	$212 \times 102 \times 54$

Table 2.1b. Properties of mortar used.

Mortar	Compressive strength (N/mm^2 28 days)	Bending strength (N/mm^2 28 days)	Cement:lime:sand Volume parts	Parts by weight
A	8.2	2.2	1:1:6	1:0.48:6.72
B	3.0	0.7	1:2:9	1:0.96:10.1
C	16.1	3.4	1:½:4½	1:0.24:5.04
D	44.5	6.0	1:½:1½	1:0.24:1.68
E	3.1	1.0	1:½:9	1:0.24:10.1
Bolidt	50.0 (20 days)	46.0 (20 days)	–	–

Table 2.2. The material combination used in the various tests. C = compression tests, T = tension tests, S = shear tests.

Units	Mortar A 1:1:6	B 1:2:9	C 1:½:4½	D 1:½:1½	E 1:½:9
JB		D	D	D	
JG		D, T, S	D, T, S	D	D
VE		D, T, S	D, T, S	D	D
CS	D, T	D, T, S	D, T	D	D

unit was combined with two or more types of mortar. Table 2.1a shows some properties of these materials.

The flexural and compressive strength values of mortars as shown in Table 2.1b are the averages of the values that were found in the various mortars batches used throughout the programme.

The material combinations used in the tests are shown in Table 2.2.

At the time when the specimen using calcium silicate units had to be made for Series 1 of the tension and compressive tests, there was no mortar C 1:½:4½ by volume (b.v.) available. That is why mortar A 1:1:6 (b.v.) was used.

2.2.2 *Manufacturing of specimens*

With the compression, tension and shear tests three different types of specimens were used:

   – Mortar prisms;
   – Unit prisms;
   – Masonry specimen.

With the analysis of the test results [9-11] it became apparent that the properties of the mortar prisms are not representative for the properties of the mortar in the joint. As a consequence these tests will not be discussed here.

The mortar specimens were prisms of $40 \times 40 \times 160$ mm. This is the standard size that is used for qualification tests according to NEN 3835 'Mortars for masonry with units or blocks using clay units, calcium silicate units, concrete and autoclaved aerated concrete'. The geometry of the specimen is described in Sections 2.3-2.5 for respectively the compression, tension and shear tests.

Of all the specimens in which units were used, the loaded surfaces consisted of unit on the outside, except those of Series 2 of the compression tests. The loaded paces of clay units were made smooth and level by grinding with water. With the Vijf Eiken units the brand mark was ground off as a result of which the unit became 4 mm thinner. In the compression tests Series 2 the loaded surfaces consisted of a mortar joint.

For the compression tests, specimens with 0.3 mm thin 'Bolidt' joints were also made. Bolidt is a two component grout. The grout mortar was processed according to the manufacturer's prescription. The clay units were made smooth and level by grinding on either side. The specimens were made by spreading a thin layer of Bolidt on the contact surfaces of the dry polished units by means of a spatula. In each case two units were put on top of each other and after about 1 hour three pairs of units and one loose unit were recombined to form one prism of seven units. The thickness of the layer of the two component material between the units was 0.3 mm on average. The specimens with Bolidt joints were tested within 24 hours after they were made.

For the deformation-controlled tension tests half units and unit prisms were used. The half units needed for the masonry specimen with mortar were cut out of the centre of a unit. The same applied for the tensile bond tests that were carried out simultaneously with the shear tests. The necessary unit prisms were also cut out symmetrically from the centre of a unit. In the unit prisms a notch was cut. Cylindrically-shaped clay unit specimens were made as well. These were obtained by taking cores out of unit by means of a water-cooled tubular drill. Both the upper and lower side of the specimen were ground. A notch was cut all round.

The joint thickness of the masonry specimen varied between 12 and 15 mm in the different test series. The units were laid at the correct height using a cord guiding as is common practice in The Netherlands. The joints were filled completely and finished with a pointing trowel.

After laying the units a weight was put on top of each masonry so that each masonry had an average compressive stress of 0.002 N/mm². Eurocode 6 (EC6, [5]) prescribes a pre-pressure of 0.001-0.0012 N/mm² for small test walls. After that the masonry's were closely covered with a polyethylene foil for three days and were subsequently placed into a room with a constant climate of 20°C/60% (RM). To obtain plan-parallel shear masonry's (consisting of two units and one joint) four adjustable supports with a height of two unit thickness + joint thickness were fixated on a table. The specimen was placed between the four supports. A perspex plate was pressed on top of the upper unit as far as the supports.

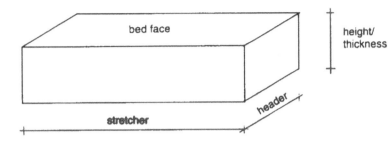

Figure 2.1. Terminology.

The pre-treatment of units used in specimens with the compression and tension tests differed from that of the shear tests [9, 11]. The immersion time for the compression and tension tests is held uniform at 7.5 minutes for all the units. For the shear tests specimen shorter immersion times, differing per unit, were used. These times were a compromise between those necessary for an optimum suction rate and moisture content. With the shear tests this resulted in a better bond for the clay units and a worse bond for the calcium silicate units.

2.2.3 *Terminology*

For the indication of dimensions and directions in a specimen, the terminology as shown in Figure 2.1 will be used.

If a specimen is loaded in the direction of the largest dimension of the unit, for example, then the term loaded in the 'stretcher direction' will be used.

2.2.4 *Coding*

The specimen will be indicated by a code. As stated before, three different types of specimens were used:
– Specimen consisting of mortar only (mortar prisms);
– Specimen consisting of units only;
– Specimen consisting of units separated by joints.
The code is in principle set up as follows:
– Test, unit and joint type.
The test type can be: C = compression test, T = tension test, S = shear test, and TS = accompanying tension test with shear test.

For the unit type and joint type the codes shown in Table 2.1 are used. With the tension tests, the unit type is also referred to by the indications VE-r and JG-r. These indicate the cylindrically-shaped (round) specimen that were drilled out of one unit. The mortars and their recipe are also given in volume ratios. With the tension tests on prisms and cylinders made out of one unit, the code for the joint has been omitted.

2.3 MECHANICAL PROPERTIES OF MASONRY IN COMPRESSION

This section discusses the results of two series of deformation-controlled compression tests on masonry. The specimens were manufactured and tested in the Pieter van

Musschenbroek Laboratorium of the BKO department of the Eindhoven University of Technology.

The research was co-ordinated, interpreted and reported by the Eindhoven University of Technology [9, 10].

With the first series, 15 stack bonded prisms made out 5 units laid on top of each other were tested. The second series consisted of 30 wide and 30 narrow specimens. Wide and narrow specimen of the same type of masonry were made on the same day and tested on the same day. The narrow specimens were almost the same as those of the first series, the wide specimens had dimensions according to NEN-EN 1052-1 (the concept version of May 1991 [23]). They were two units wide. The aim of this series of tests was to find a relationship between the narrow specimens of Series 1 and the wider specimens according to NEN-EN 1052-1, to be able to place the results of the narrow specimens in an international framework.

The influence of the mortar and unit strength on the strength of the masonry has been mainly investigated in the second test series by using combinations of four mortar qualities and four unit qualities.

All combinations were carried out in triplicate.

To investigate the behaviour of the units without the influence of uncertain joints in compression, two specimen of each type of unit were made and tested. These specimens consisted of 7 smoothly grounded units bound to each other with a very thin layer of Bolidt (≈ 0.3 mm).

2.3.1 *Specimens*

The dimensions of the specimen of Series 1 were chosen in such a way that a uniaxial stress is present in the centre part of the specimen and eccentricities do not play an important role. For the remainder the specimen has been kept as simple as possible. The cross-section of the specimen is equal to the unit's bed face. The height was 320 mm in Series 1 and 340 mm in Series 2. The difference in height was caused by 2 extra joints that were added to the upper and lower side in Series 2. The slenderness of the specimen amounts to 1.6 or 3.2 depending whether the stretcher or header direction was taken as the thickness. The dimensions of the units were approximately $210 \times 100 \times 50$ mm. The dimensions of the narrow specimen and of the specimens according to NEN-EN 1052-1 are shown in Figure 2.2.

Figure 2.2. Dimensions of the masonry compression specimens.

The length of the wide specimens is according to the concept version of NEN-EN 1052-1 of May 1991 [23] determined by the length of two complete units and the thickness of 1 header joint. The length of the header has been adapted accordingly. The approximate dimensions for the wide specimens were 430 × 100 × 340 mm (length × width × height). For the narrow specimens the dimensions were 200 × 100 × 320 mm for Series 1 and 200 × 100 × 340 mm for Series 2.

2.3.2 *Testing*

The apparatus that was used for all the compression tests, is an electrically-driven hydraulic press, made by Schenk Trebel, with a maximum compression of 2.5 MN. The jack was deformation-controlled. The displacement speed of the jack was about 28 μm per min. The maximum load was reached in 15 to 20 minutes. For specimen with VE unit the load speed used is about 0.5 N/mm²/min.

The influence on the behaviour of a test piece by interface layers such as sliding membrane, cardboard or the like, placed between the test piece and the loading plates is not clear, therefore these devices were not used and the test remained as simple as possible. This is also of practical importance for the determination of standard parameters with simple tests.

The position of the LVDT'S to measure the reduction in length of the specimen is shown diagrammatically in Figure 2.3.

The following measurements were taken for each specimen:

a) The reduction in length and the height of specimen, by registration of the displacement of the corners of the compression plate with long LVDTs (Linear Variable Differential Transducer, this is an inductive transducer).

b) The reduction in length of the centre part of the wide test piece over a gauge length of about 112 mm in the centre of the specimen, with LVDTs. In this measuring zone two joints occur.

c) The deformation of the units with clips with strain gauges. The gauge length was 40 mm.

d) The deformation in stretchers direction perpendicular to the load direction with LVDTs over 150 mm.

Figure 2.3. Position of the transducers on the specimens for the compression test.

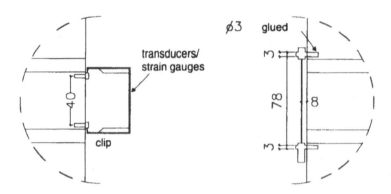

Figure 2.4. Attachment of the transducers to the compression specimens.

The chosen measuring length for measurement type b is in conformity with [23]. The measurements as stated before, except for measurement a, have not been carried out with all specimens. For a detailed survey refer to [9, 10].

The LVDTs for the measurements were fixed to the specimen with brassthreaded, 3 mm round bars. These bars were glued into a drill hole of about 10 mm depth. The transducers were placed as close as possible to the surface of the specimen (refer to Fig. 2.4). That way possible measuring errors due to the rotation of the bars are smaller.

2.3.3 *Progress of the test*

The upper loading plate hangs on a self locking ball. With a load of 12 kN the gap between the loading plate and this ball hinge is closed. At that moment the loading plate rests, practically without friction in the ball hinge, on the specimen and will position itself. The upper side of the specimen and the loading plate are then parallel to each other. After a start-up phase up to about 20 kN (0.5 or 1 N/mm^2) the friction in the ball joint will increase. Measurement a. shows that the displacements of the corners of the upper loading plate stay more or less the same with increased loading. From that it can be concluded that the eccentricity of the load on the specimens remained the same during the further development of the test.

Each specimen was loaded beforehand to about 50 kN (1.25 or 2.5 N/mm^2) and subsequently the load was reduced to zero. This was done to check the whole set-up and to position the specimen and the loading plate in relation to each other.

The development of cracks cannot easily be determined. At about 80 to 85% of the fracture load the crack formation started to manifest itself, often only audible at first, and only later on visible.

Especially with the harder Joosten units one could hear some cracking when the crack development increased. With some JG specimens and most of the JB specimens crack formation appeared suddenly, sometimes accompanied by an explosive failure. With the Vijf Eiken specimens the cracking process progresses more gradually. The cracks are almost straight and run parallel to the line of action of the load. In the upper part of the top unit they sometimes run with an angle of 45° to the corners. When the load was released large flat pieces came loose with a thickness of

Table 2.3. Compressive strength values of mortar, units and masonry (each value is the average of three tests).

Mortar c:k:z (v.d.)	Mortar $f^1_{c:mortar}$ (N/mm²)	Kind of unit	7 units with Bolith $f^1_{c:unit}$ (N/mm²)	Masonry wide $f^1_{c:wide}$ (N/mm²)	Masonry small $f^1_{c:narrow}$ (N/mm²)	$f^1_{c:wide}$ - $f^1_{c:narrow}$	EC 6 f^1_{ck} (N/mm²)
Series 1							
B 1:2:9*	6.5	VE	14.2**	–	10.8	–	7.6
	6.5	JG	35.8**	–	20.6	–	12.1
	6.5	CS	30.0**	–	19.4	–	8.0
C 1:½:4½*	26.6	VE	14.2**	–	10.2	–	10.8
	26.6	JG	35.8**	–	19.2	–	17.2
Series 2							
E 1:½:9	2.9	VE	14.2	7.1	6.5	1.10	6.2
	2.9	JG	35.8	8.6	9.0	0.96	9.9
C 1:½:4½	9.8	VE	14.2	10.4	8.7	1.20	8.4
	9.8	JG	35.8	15.9	17.6	0.98	13.4
	10.3	JB	64.8	20.9	20.9	1.11	20.0
D 1:½:1½	47.9	VE	14.2	12.8	10.5	1.22	12.5
	47.9	JG	35.8	36.9	31.2	1.18	19.9
	41.0	JB	64.8	45.5	39.8	1.14	28.3
A 1:1:6	10.4	CS	30.0	19.6	20.1	0.98	9.0
C 1:½:4½	6.3	CS	30.0	20.2	20.2	1.00	7.9

*Tested together with the specimen of masonry (110 or 200 days). **Average of two test results.

15 to 20 mm, often from several units on top of each other at the same time. With further deformation the specimen falls apart and a diabolo-shaped part remains.

The tests on masonry specimen made out of calcium silicate units showed a similar result as the clay unit specimens. Here too, the process developed more gradually than with the JB units.

2.3.4 *Compressive strength*

The determined compressive strength values of masonry are shown in Table 2.3. Furthermore, the corresponding bending strength and compressive strength values of the mortars have been determined according to NEN 3835. The compressive strength of the units in conformity with NEN 3836 is shown in Table 2.1a in Section 2.2.1. In Table 2.3 the compressive strength of the prisms with Bolidt joints is also shown. Table 2.3 also gives the characteristic masonry compressive strength values obtained when the values of the mortar and unit compressive strength values according to respectively NEN 3835 and NEN-EN 772-1 [26] are substituted in the following equation taken from EC 6 [5]:

$$f_{ck} = 0.6 \cdot (f_{c; unit})^{0.65} \cdot (f_{c; mortar})^{0.25} \tag{2.1}$$

EC 6 does not mention how the mortar compressive strength should be determined. The assumption is that there will be a reference to NEN-EN 1015. NEN-EN 1015, as far as the determination of the compressive strength is concerned, nearly equals NEN

3835. The unit compressive strength according to NEN-EN 772-1 is for the tested unit size unit lower by a factor of approximately 0.75 than the value according to NEN 3836. Therefore, the values of Table 2.1a multiplied by 0.75 were substituted in Equation (2.1). The characteristic values calculated by means of Equation (2.1) are shown in Figure 2.5 and are compared to the averaged measured values. For a good comparison the 'ideal' relation between the measured and calculated values is also shown. The characteristic value implies:

$$f'_{ck} = f'_c (1 - 1.64v) \tag{2.2}$$

With an assumed coefficient of variation $v = 20\%$, the ideal relation between calculated characteristic strength and measured strength is: $f_{ck} = 0.67 \cdot f_c$.

The EC 6 equation being a formula for the design of structures should never overestimates the real strength. This appears not to be the case. The strength of masonry is defined by more factors than only the unit and mortar strength.

Other factors are:
– Moisture condition of the unit;
– Finishing;
– Joint width;
– Suction rate of the unit;
– Dimensions of the unit (ratio between joint thickness and unit height);
– Inner cracks and stresses within the unit;
– Craftsmanship of mason;
– Filling of the joints;
– Finishing of the joints.

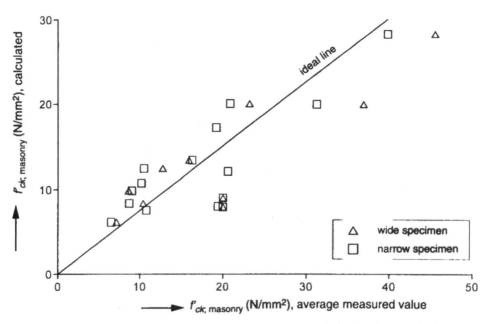

Figure 2.5. Comparison of measured and calculated masonry compressive strength values according to EC 6.

Since the properties of a mortar are influenced by the units between which it hardens, it is principally incorrect to take the strength of the 'corresponding' mortar prism as a starting point to determine the compressive strength of the masonry.

Stronger mortars, however, do give stronger masonry which comes up to the expectation (refer to Fig. 2.6). The dotted curves indicate this tendency.

Per series of 3 similar specimens the age of the specimen does not significantly influence the strength because the age differs 4 days at the most. The specimen of Series 1 were about 100 and 200 days old, those of Series 2 were 32 days old on average.

For the tests of Series 2 the strength development of the mortar was monitored in the course of time. Based on this it can be said that between 28 and 200 days there is hardly an increase in compressive strength. This tendency is confirmed in references [11, 15 and 16].

It should be noted that the strength of the prefab mortars from Series 1 at the moment of testing was significantly higher than what can be expected on the basis of the recipe.

Table 2.4 gives the average of the ratio of the compressive strength between wide and narrow specimen, arranged according to unit type.

In the Belgian standard NBN 24-301 [22] the following general equation is given to define the influence of the dimensions of the specimens on the strength of the specimen.

$$c = \frac{f'_x}{f'_{c20}} = 0.65 + \frac{0.7}{\left(1 = \frac{1+w}{400}\right) \cdot \left(\frac{2h}{1+w}\right)^{1.25}} \tag{2.3}$$

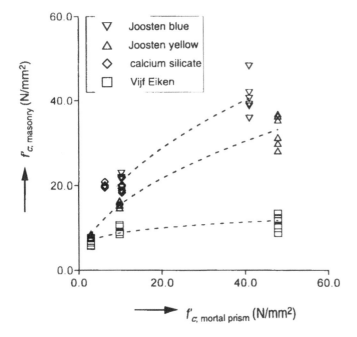

Figure 2.6. Masonry compressive strength dependent on the mortar compressive strength (NEN 3835), arranged according to unit type.

where: f'_x is the compressive strength of the specimen, f'_{c20} is the compressive strength of a cube with sides of 200 mm, $l \times w \times h$ are the dimensions of the specimen (height of the specimen measured in the loading direction.

This formula was proposed by Dutron to: 'express the general ratio between the compressive strength values of concrete measured on specimen with various shapes and dimensions'. The strength ratio calculated with this formula between a wide and a narrow specimen is 0.88/0.80 = 1.10. This is nearly equal to the experimentally found average value of 1.11. This does not imply that Dutron can be used for all specimen piece dimensions and types of material. The value of 0.99 found for the calcium silicate unit specimen contradicts Equation (2.3). Further research with other unit and mortar qualities is desirable.

The masonry's compressive strength values of the tests in Series 1 and 2, as well as their age are given in Table 2.5.

The differences found between the compressive strength values of the narrow VE.C and the JG.C specimens of Series 1 and 2 are the result of differences in mortar composition and, to a limited extent, the result of differences in age of the specimens while testing. What strikes most is that the strength values of the specimens of Series 1, made of the same unit but with different mortars, are nearly equal. Compare VE.C (10.2 N/mm²) to VE.B (10.8 N/mm²) and JG.C (19.2 N/mm²) to JG.B (20.6 N/mm²). With the specimens of Series 2 the strength values for various mortar unit combinations, however, are different. Compare, for example, from Series 2 the narrow specimen (refer to Table 2.3), VE.E (6.5 N/mm²) to VE.C (8.7 N/mm²) and

Table 2.4. Ratio between the compressive strength values of wide and narrow specimen made from clay unit.

Unit	Mortar D 1:½:1½	Mortar C 1:½:4½	Mortar E 1:½:9	Average for each kind of unit
VE	1.22	1.20	1.10	1.17
JG	1.18	0.98	0.96	1.04
JB	1.14	1.11	–	1.13
Average for each kind of mortar	1.18	1.10	1.03	1.11(8.8)*

*Coefficient of variation.

Table 2.5. Compressive strength values and age of the narrow specimens made out of masonry in Series 1 and 2.

Mortar c:k:z (v.d.)	Kind of unit	Series 1 f_c (N/mm²)	Age (days)	Series 2 f_c (N/mm²)	Age (days)
C 1:½:4½	VE	10.2	95	8.7	32
	JG	19.2	102	16.3	32
	CS	–	–	20.1	32
A 1:1:6	CS	–	–	20.1	32
B 1:2:9	VE	10.8	188		
	JG	20.6	195		
	CS	19.4	200		

VE.D (10.5 N/mm²), and JG.E (9.0 N/mm²) to JG.C (16.3 N/mm²) and JG.D (31.2 N/mm²).

2.3.5 *Elastic modulus*

Only in the case of a purely linear elastic material can be spoken of the elastic modulus. With a non-linear σ-ε diagram a choice has to be made for a defining method. Since there was a great variation in the strength of the various types of specimens, it has been decided to define the elastic modulus at a load of 35% of the determined fracture load. The measured force-deformation relation is indicated by a straight line through the origin and the point on the measured force-deformation relation where the force is 35% of the fracture load.

2.3.5.1 *Stiffness of the units*
The stiffness of the units is in the first place determined with specimen consisting of six units grounded smooth on both sides. The deformations measured on these specimens over various measuring lengths were not proportional in relation to each other. That is why it was decided to use specimens consisting of seven units made smooth and level on both sides that were bonded with Bolidt. The height of these specimens was also equal to the height of the masonry specimens. With these specimens the deformations, however, either measured over the length of the entire specimen, or measured over two joints (112 mm) and measured on a unit over 40 mm length were proportional. The measured elastic moduli of the units determined with these deformations are given in Table 2.6. This elastic modulus was also determined from the average deformation measured at 35% of the fracture load.

2.3.5.2 *Elastic moduli of specimen made out of masonry*
The elastic moduli of the wide and narrow specimen made out of masonry have been calculated with the deformations measured between the loading plates with a load of 35% of the apparent fracture load. The measuring length was equal to the specimens height which was 340 mm. The results are shown in Table 2.7.

The ratio between the stiffness of the wide and the narrow specimen varies between 0.78 and 1.05. A relation with unit or mortar properties could not be found. These large differences can be attributed to the difference in smoothness and mortar quality of the bottom and top joint and influences due to the confined deformation at the extremities of the specimen by the loading plates.

When the found elastic moduli are plotted versus the compressive strength values of the corresponding specimen then it appears that there tends to be a linear relation

Table 2.6. Elastic moduli of the units determined from stacks of 7 grounded units joined together by Bolith.

Kind of unit	E_{unit} (N/mm²)
LB	14400
JG	16700
VE	6050
CS	13400

Table 2.7. Elastic moduli of masonry determined on masonry specimens measured over a length of 112 mm, and between the loading plates (measuring length 340 mm). Each value is the average of three tests.

Mortar c:k:z	Kind of unit	$E_{specimen}$ measuring length 340 mm (N/mm^2)	$E_{masonry}$ measuring length 112 mm (N/mm^2)	$\dfrac{E_{specimen}}{E_{masonry}}$
Wide specimens				
E 1:½:9	VE	2140	2820	0.76
	JG	2380	5590	0.43
A 1:1:6	CS	6050	7870	0.77
C 1:½:4½	VE	4100	3890	1.05
	JG	7710	9920	0.78
	JB	7520	10250	0.73
	CS	6250	9240	0.68
D 1:½:1½	VE	3160	4240	0.74
	JG	14230	15470	0.92
	JB	11360	13250	0.86
Narrow specimens				
B 1:2:9	VE	3950*	4210	0.94
	JG	13030*	15880	0.82
	CS	6330*	6710	0.94
A 1:1:6	CS	6130	7670	0.78
C 1:½:4½	VE	3870*	**	–
	JG	13330*	13200	1.01
	CS	5350	8980	0.71

*Values of Series 1: The contact area between specimen and test arrangement consisted of grained units in contrast with series where two extra joints the contact area formed. **No reliable values were obtained, in consequence of the test has been modified.

between strength and elastic modulus of the specimen. Some influence of the type of unit can be noticed with the strongest specimens.

With the tested calcium silicate unit specimen both the strength and the elastic modulus are nearly independent of the width of the specimen. The elastic moduli that are discussed here were determined with a load of 35% of the fracture load, so for each type of specimen at another stress level. This mainly influences VE unit because their load-deformation relation is the least linear.

Elastic modulus of masonry, determined with measurements over 112 mm. The deformations in the centre of the specimen are measured with LVDTs over a length of 112 mm (refer to Table 2.7). In this measuring field there were two joints present. Figure 2.7, for example, shows the measured force-deformation relation of two joints and units for a VE specimen. This picture is representative for all the tests.

The compressive strains in the centre areas is noticeably smaller then that of the entire specimen. This appears from the comparison of the results of measurements over the length of the entire specimen with measurements over 112 mm. A comparison of the determined elastic moduli is also given in Table 2.7.

The differences result from the fact that the upper and lower side of the specimen are not absolutely smooth, and because of differences in the stress distribution be-

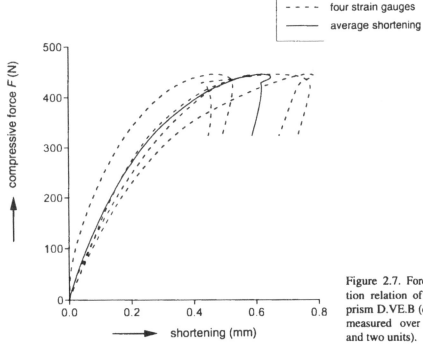

Figure 2.7. Force-deformation relation of a masonry prism D.VE.B (deformation measured over two joints and two units).

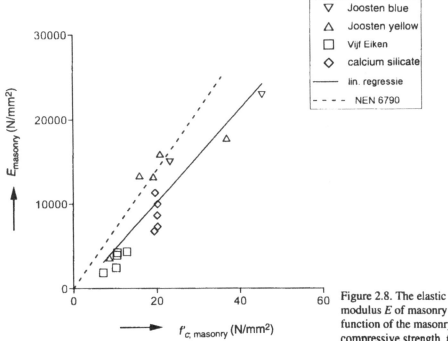

Figure 2.8. The elastic modulus E of masonry as a function of the masonry compressive strength f'_r.

tween the parts at the extremities of the specimen and that in the centre. Furthermore, the combination JG with mortar C 1:½:4½ shows a striking difference with the narrow specimen of Series 1 (E_{specimen} = 13200 N/mm^2) and Series 2 (E_{specimen}) = 9920 N/mm^2). With the narrow specimen of Series 1 the prefab mortar with the high compressive strength was used which probably caused the high values that are found here. In Figure 2.8 the elastic moduli of the masonry (determined with the measurements over 112 mm for both the wide and the narrow specimens) are plotted against the corresponding compressive strength values. In contrast to the values of the elastic moduli with 35% of the fracture load shown in Table 2.7, the elastic modulus shown in Figure 2.8 was determined at a stress of 4 N/mm^2 for all specimens.

Suppose there is a rectilinear relation, then the following relation can be determined with the aid of linear regression:

$$E_{\text{masonry}} = 526 \cdot f'_{c;\text{masonry}} - 568 \text{ N} / \text{mm}^2 \tag{2.4}$$

The linear regression was carried out on the 12 averages of the observations of the small groups of 3 similar specimens from Series 2. The correlation coefficient r was 0.89. If the 5 results of Series 1 are also taken into account then $r = 0.80$. The factors in Equation (2.4) change only slightly then, also because the number of data in Series 1 is limited in comparison to Series 2.

Considering the limited number of observations for each type of unit the given relation should only be seen as an indication.

To make certain whether there is either a general relationship or a relationship per type of unit, it is recommendable to work with more unit and mortar qualities in future research. In NEN 6790 TGB unit structures it is assumed that:

$$E_{\text{masonry}} = 1000 \cdot f_{c;\text{ rep; masonry}} \tag{2.5}$$

With regard to the average compressive strength this is approximately equal to:

$$E_{\text{masonry}} = 700 \cdot f_{c;\text{ av; masonry}} \tag{2.6}$$

Equation (2.6) is also shown in Figure 2.8. For the tested combinations it appears that NEN 6790 overestimates the elastic modulus somewhat. It should be noted that this conclusion depends on the measuring method of 'the' elastic modulus and on the method with which the compressive strength has been determined.

The modulus of elasticity of the mortar. Table 2.8 shows the moduli of elasticity of the mortars in the masonry specimen. These elastic moduli have been calculated by means of the measured deformations over 112 mm and the elastic moduli of the units determined by the Bolidt tests.

In Series 2 extremes were used for the mortar composition and the unit quality which is reflected in the values determined for the elastic moduli. The high values of the elastic moduli of mortars between Joosten units from Series 1 are striking, compare the values of JG.B and JG.C of Series 1 with JG.C (wide) of Series 2. Apart from Series 1 the elastic modulus with stronger mortars appears to be higher. Compare, for example, the results of VE.E with that of VE.C and VE.D. The elastic moduli found for the mortars that were designated in the Pieter van Musschenbroek laboratory turned on our own account, and which were used for wide and narrow calcium silicate unit specimen are not significantly different, refer to the last two

Table 2.8. Elastic moduli of mortar in wide masonry specimens, calculated by means of the measured deformations over 112 mm and the elastic moduli of the units determined with Bolith tests.

Mortar	Unit	E_{mortar} (N/mm^2)						
		VE wide	VE narrow	JG wide	JG wide	JB narrow	CS wide	CS narrow
E 1:½:9		1090		1770				
B 1:2:9			2860*		14900*			3350
A 1:1:6							3690	3530
C 1: ½:4½		1950	1410*	4510	13300*	5320	4990	4700
D 1:½:1½		2120		12010		9330		

*Values from Series 1 (prefabricated mortars).

columns of Table 2.8. This also applies to similar mortars that were combined with JG and JB units, refer to JG.C and JB.C.

The units influence the stiffness of the mortar, both with test Series 1 and with Series 2. The elastic modulus of the mortar is larger when the compressive strength of the units is larger, compare VE.C with JG.C, VE.E with JG.E and VE.D with JG.D. This is partly caused by the generally smaller suction capacity of the stronger units, as a result of which the mortar can harden under better conditions, refer to [14].

A remark should be made about the method by which the mortar's moduli of elasticity were determined. Deformation in the units within the gauge length were calculated using the modulus of elasticity obtained with the 'Bolidt' test (Section 2.3.5.1). These deformations were substracted from the measured deformations. The remaining deformations were attributed to the mortar joint within the gauge length. This implies that measurements errors and deviations in stiffness of the units influence the presented joint stiffness to an unknown extent and cause an unknown extra variability.

Although the specification in volumes of the mortar composition is equal for both series of tests, there were some differences. The mortar for the tests of Series 1 was factory made in the ready mixed form. The mortar for the tests of Series 2 was designated in the laboratory. Different values were used ranging from 1450-1600 kg/m^3.

2.3.6 *Transverse contraction*

The deformation vertical to the loading direction measured at the middle unit over a length of about 150 mm. Since the measurable deformations develop under widely varying loads, it is not easy to calculate a Poisson's ratio v by means of the deformations measured on a single unit. It is remarkable that in the beginning the deformation in the stretcher direction is very small. This corresponds with the theoretical examination by Beranek [25]. Just as with the measurements of the unit's deformation parallel to loading direction, these measurements show that the unit's deformation differs from that which one might expect on the basis of stress distribution.

If the deformation in the direction of the thickness is calculated with the unit's modulus of elasticity, which was determined by means of the Bolidt tests instead of the deformation on the mortar specimens, then we obtain the results as shown in Table 2.9. Table 2.9 also shows the values for v that were defined on the basis of the Bolidt tests.

In view of the slight difference between the masonry specimens and the Bolidt specimens, it appears that at the stress level at which v has been defined (maximum 50% of the fracture load) the units are not (yet) pulled apart by the mortar. In the wide calcium silicate unit specimen measurements were also taken in transverse direction. A vertical joint was present in the measuring area as well. Due to the presence of vertical joints in the wide specimen, the measured force-deformation lines show jumps and irregularities. In a few cases these lines were almost vertical, which implied that measured deformation in the direction perpendicular to the compression direction does not increase under an increasing force. Perhaps the unit could deform freely in the direction of the badly filled vertical joint, as a result of which the total deformation over the measuring length did not increase.

2.4 TENSION TESTS

This section discusses the deformation-controlled tension tests on units and masonry. The specimen were manufactured simultaneously with the compression specimen of Series 1 at the Pieter van Muschenbroek laboratory. The experimental part of this research was carried out at the Stevin laboratory of the Delft UT [12]. The research was co-ordinated, interpreted and reported by TNO [9].

The main purpose of the tests was to determine the stress-crack width diagram and the corresponding fracture energy. These data are needed for material models which are used with numerical simulations of masonry. The stress-crack width diagram shows the behaviour of unit-like material under tension (see Fig. 2.9). The surface

Table 2.9. Average Poisson's ratio v of units. a) In masonry specimens with mortar 1:2:9, b) In specimens using Bolith.

Kind of unit	E_{unit} (N/mm^2)	v (–)	
		a	b
VE	6050	0.14	0.14
JG	16700	0.28	0.28
JB	15400	–	0.19
CS	13400	0.17	0.20

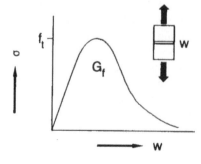

Figure 2.9. Relation between the stress-crack width diagram and the fracture energy.

a

b

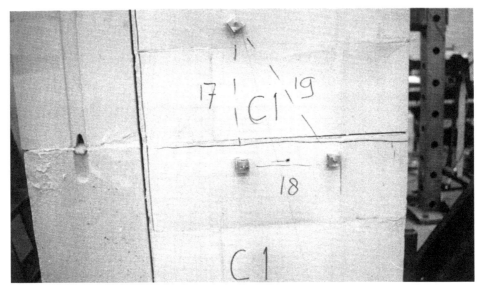

Figure 7.3. Experimental observations [57]. a) Toothed connection that fails in pier, mechanism (2) in Figure 7.2, b) Vertical line joint connection that fails in vertical line joint, mechanism (3) in Figure 7.2.

2.4.2 *Testing*

The test arrangement of the Stevin laboratory of the Delft UT is described extensively in [13]. Here the arrangement will be discussed briefly. Figure 2.11 shows a diagram of the arrangement.

The specimen is glued between two platens in the test arrangement. The platens are kept parallel by a stiff guiding system. Four load cells were placed at the bottom of the lower plate by means of which the force in the specimens can be registered. Due to the position of the pressure boxes, the registered force is not influenced by the friction in the guiding system. The deformation speed is measured and guided by means of LVDTs on the specimen. When using LVDTs, the location of the crack has to be within the gauge length. That is why the unit prisms and cylinders were provided with a notch. With the masonry specimen the location of the cracks is determined by the joints. When the tensile strength has been reached, a crack appears locally in the area of the measuring length, while part of the specimen remains elastic. Where the crack occurs, the material loses a considerable amount of its stiffness. The deformation increases around the crack and it decreases in the uncracked part. Due to this rearrangement of deformations within the gauge length, the crack width increases more than is registered over the measuring length. Therefore, a short measuring length has a positive effect on the success of a deformation-controlled tension test.

The position of the transducers on the specimen is shown diagrammatically in Figure 2.10. The exact locations are shown in [11]. Four transducers were placed at two opposite sides of the prismatic unit specimens. On the cylinder-shaped specimen four transducers were placed equally divided on the outside perimeter. At two

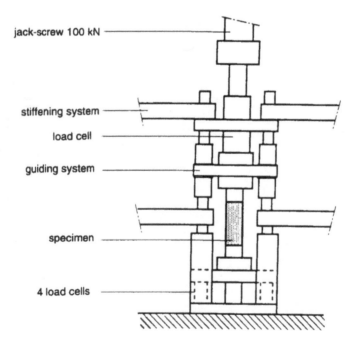

jack-screw 100 kN

stiffening system

load cell

guiding system

specimen

4 load cells

Figure 2.11. Tension test arrangement used at the Delft UT Stevin laboratory (extracted from [13]).

opposite sides of the masonry specimen which consisted of two half unit and one joint, in total four transducers were placed across the joint. In the case of the masonry specimen consisting of three half units and two joints, one transducer was placed on each side across two joints and the middle unit, and one transducer across the lower or upper joint. In total eight transducers were placed on this type of specimen.

2.4.3 *Tensile strength of the joints and the units*

The clay units were tested both in stretcher direction and in thickness direction, loaded perpendicular to the bed face, because they were expected to show a different behaviour in these directions. These tests did not take place on calcium silicate units. Based on the manufacturing process, an isotropic behaviour can be expected. The average tensile bond strength values are shown in Table 2.10. Besides, the variation coefficients (v.c.) are shown. It should be emphasized however, that the v.c. value is not very reliable, due to the limited number of test results.

In most of the masonry specimens a crack developed in the interface between the joint and the unit. In one specimen from the T.JG.C series a crack developed in the joint itself. In the masonry specimen with two joints the crack developed four times in the upper joint and five times in the lower joint. The cracked bond interface of the calcium silicate units was smooth.

From the results as shown in Table 2.10 the following can be concluded:

– The Joosten unit is at its strongest in the thickness direction (cylinder), whereas the Vijf Eiken unit is at its strongest in the stretcher direction (prism). This is attributed to the difference in direction of the layered structure of the units resulting from a different manufacturing process (mould versus extrusion process).

– The bond tensile strength of T.VE.B. (0.22 N/mm^2) is significantly higher than that of T.VE.C. (0.13 N/mm^2). This is remarkable because mortar C (1:½:4½) is stronger than mortar B (1:2:9).

– An increasing mortar strength (B-A-C) only has a positive effect on the tensile bond strength when combined with the Joosten unit. This is one of the reasons why it was concluded that a mortar prism is not representative for the mortar that hardens between the units.

– There is a wide range of test results. This is caused by the large number of factors that play a role in the development of bond [14]. An influence that is identified in this research is the real interface. From a visual inspection of the cracked area it

Table 2.10. Average (bond) tensile bond strength.

Tensile strength				Tensile bond strength			
Type of specimen	f_{tu} (N/mm^2)	v.c. (%)	Amount of tests	Type of specimen	f_{tu} (N/mm^2)	v.c. (%)	Amount of tests
JG prism	2.36	21	2	T.JG.B	0.30	24	3
JG-r cylinder	3.51	3	3	T.JG.C	0.50	29	6
VE prism	2.47	14	3	T.VE.B	0.22	45	3
VE-r cylinder	1.50	4	3	T.VE.C	0.13	100	3
CS prism	2.34	10	3	T.CS.B	0.29	34	4
				T.CS.A	0.33	51	5

appeared that the interface of the different specimen varied extensively and could be considerably smaller than the gross cross sectional area of a specimen. This phenomenon will be discussed in Section 2.4.6.

2.4.4 *Elastic moduli of the joints in tension*

On the basis of measurements on the masonry specimen it is possible to calculate the elastic modulus of the joints. The gauge lengths, however, 'include' pieces of unit as well. To determine the deformation in the joint itself, the measurements had to be corrected. For this purpose the elastic moduli of the units were used, based on the compression tests on small stacks of units with gapfilling of Bolidt as described in Section 2.3. The elastic moduli in tension, after correction of the measurements were up to half of the collapse load by means of linear regression. The correlation coefficient for all determinations was larger than 0.97. Up to approximately 80% of the collapse load, the specimens show a linear behaviour. The secant elastic modulus of the mortar joints at the collapse load amounts to 50 to 70% of the values shown in Table 2.11. From the results it can be concluded that for the elastic modulus of the mortar joint the same tendencies are present with the bond tensile strength. The difference between the values obtained here and the values resulting from the compression tests is remarkable. We will refer to this in Section 2.6.

2.4.5 *Fracture energy and 'post peak' behaviour*

The fracture energy is defined in Figure 2.9. Figure 2.12 shows the stress-strain diagrams of the T.KZ series as examples of the coarse of the deformation-controlled tension tests.

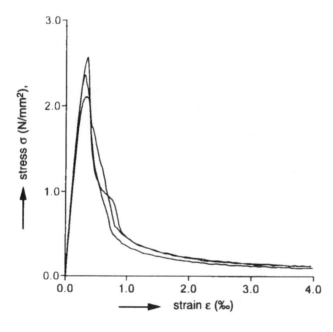

Figure 2.12. Development of deformation-controlled tension tests (T.CS series).

Table 2.11. Elastic modulus of the joints in tension.

Type of specimen	E (N/mm^2)	v.c. (%)	Amount of tests
T.JG.B	2970	9	3
T.JG.C	6035	20	6
T.VE.B	700	22	5
T.VE.C	666	69	3
T.CS.B	2544	23	4
T.CS.C	2630	22	5

Table 2.12. Fracture energy.

Units				Mortar joint			
Type of specimen	G_r (J/m^2)	v.c. (%)	Amount of tests	Type of specimen	G_t (J/m^2)	v.c. (%)	Amount of tests
JG prism	117	–	1	T.JG.B	12	64	3
JG-r cylinder	128	3	3	T.JG.C	7	53	6
VE prism	61	24	3				
VE-r cylinder	73	3	3	T.VE.B	8	66	4
CS prism	67	14	3	T.VE.C	4	43	3

Table 2.12 shows the fracture energy values of the units and of the bonding surface in the masonry specimens as derived from the load-displacement diagrams. The masonry specimen made out of calcium silicate units showed uncontrolled failure after the bond tensile strength had been reached. That is why the table does not show the series T.CS.A and T.CS.B. The uncontrolled failure of these specimen does not mean that the fracture energy is zero. It can be expected that the fracture energy of the bond interface is smaller with calcium silicate units than with clay units because interlock will play a smaller role during the crack formation in calcium silicate units according to the extremely smooth fracture area.

The fracture energy of the bonding surface is very low in comparison with the units, which is in conformity with a relatively brittle behaviour, but after the tensile strength value has been reached the deformation can still increase without the tension dropping straight away to zero again. The fracture energy of the units is about ten times as high as that of the bonding surface.

With regard to the 'post peak' behaviour not only the magnitude of the fracture energy is of read importance, but also the shape of the descending curve. The shape of the descending curve for concrete can be described by means of [13]:

$$\frac{\sigma}{f_{tu}} = \left\{ 1 + \left(c_1 \frac{w}{w_c} \right)^3 \right\} \cdot e^{-c_2 \frac{w}{w_c}} - \frac{w}{w_c} \cdot (1 + c_1^3) \cdot e^{-c_2} \tag{2.7}$$

where: $c_1 = 3$, $c_2 = 6.93$, $w =$ is the crack width, $w_c =$ is the crack width at which no stress is transferred any more, for which the following applies:

$$w_c = 5.14 \frac{G_f}{f_{tu}} \tag{2.8}$$

When the test results are compared with Equation (2.7) one can conclude that this formula can well be used for the units and for the joints (bonding surface). This applies to both the individual test results and the average values per series.

2.4.6 *Influence of the real (net) bonding surface*

By investigating the area of adhesion of the masonry specimens after fracture, it appeared that the surface area of adhesion could be considerably smaller than the cross sectional area of the specimen. The net bonding surface was determined by 'looking' where bonding had taken place. Naturally, the value of such a determination is limited, although possible tendencies van be observed. Figure 2.13 shows an example of the determination.

The average net bonding surface of the specimen amounted to 35% of the cross section of the specimen. Assuming that the net bonding surface is rectangular and that the reduction of the area of adhesion is mainly caused by the presence of edges (shrinkage), then the bonding surface of a continuous half unit wall amounts to 57% of the cross sectional area. The fracture energy of the wall is then 1.7 times as large as that of the specimens. The same holds true for the tensile strength. It should be noted that the influence of the header joints is ignored.

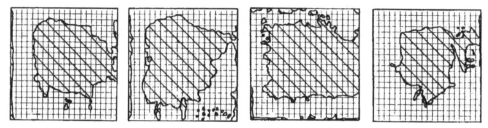

Figure 2.13. Net bonding surface of the T.VE.B specimen.

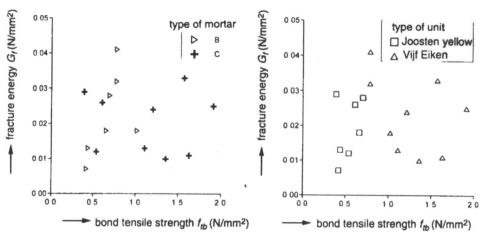

Figure 2.14. Fracture energy of the real area of adhesion in relation to the bond tensile strength, arranged according to type of mortar and type of unit.

Table 2.13. Fracture energy and tensile bond strength based on gross cross sectional area (gross) and the nett area of adhesion (net).

Type of specimen	Tensile bond strength		Fracture energy	
	$f_{th;gross}$ (N/mm^2)	$f_{th;net}$ (N/mm^2)	$G_{f;gross}$ (J/m^2)	$G_{f;net}$ (J/m^2)
T.VE.B	0.26 (45)*	0.56 (26)	8 (66)	17 (54)
T.VE.C	0.13 (43)	0.57 (51)	4 (83)	22 (40)
T.JG.B	0.30 (24)	0.58(16)	12 (64)	30 (38)
T.JG.C	0.50 (29)	1.47 (20)	7 (53)	10 (48)

*Coefficient of variation in brackets

When the net bonding surface is taken into account in the analysis, the coefficient of variation of tensile strength values and fracture energy reduces. Table 2.13 shows the influence of the bonding surface. Section 2.4.3 focused on the high tensile bond strength value of T.VE.B in comparison with that of T.VE.C. From Table 2.13 it can be observed appears that the bond strength values of these specimens do not significantly differ when they are based on the net area of adhesion.

In Figure 2.14 the fracture energy of each joint has been plotted against the tensile bond strength based on the net area of adhesion.

Based on Figure 2.14, it was concluded that it is impossible to define a relationship between the tensile bond strength and the fracture energy.

Assuming that the fracture energy is independent of the units and mortar that were used, then the average fracture energy of the specimens amounts to 7.4 J/m^2, and 12.5 J/m^2 for a continuous wall. A wide coefficient of variation should be taken into account.

2.5 JOINT SHEAR TESTS

This section describes the experimental research of the material behaviour of joint and bonding surface in shear, under the influence of various compressive stress levels perpendicular to the shear plane (the joint). This part of the research was carried out at TNO. Each of the six material combinations used, was tested in triplicate at three different compressive stress levels.

To determine a possible relationship between the bond tensile strength and the bond shear strength, uniaxial tension tests were also carried out with the shear tests on specimens that had been manufactured at the same time. Besides, CEN shear tests ('triplet test') were carried out, by means of which the shear research in hand can be placed in an international framework.

In the shear tests two units are shifted in stretcher direction in relation to each other. The relative displacement of the units is referred to as 'relative shear displacement' but also 'shear displacement' for short.

The behaviour in the shear tests is shown in the load shear displacement diagram of Figure 2.15.

Based on the obtained test results the following properties have been determined:
– The strength t_u at the maximum load dependent on the compressive stress;
– The shear stiffness G of the joint;

– The friction coefficient μ after reading the maximum load in the diagram;
– The so-called Mode-II fracture energy G_{fII} and the corresponding shear displacement range over which the cohesion reduces down to zero;
– The angle of dilatancy ψ between the relative transverse displacement u and the relative shear displacement v.

The properties are shown diagrammatically in Figure 2.15.

The analyses that were carried out are based on a uniform stress distribution over the entire shear plane.

2.5.1 *Specimen*

Figure 2.16 shows the masonry specimens for the shear tests.

When all specimen are examined together, the age at testing varied between 116 and 179 days. The age of each specimen is shown in [11]. It was assumed that the strength development of the mortar prism is similar to that of the joint. Considering the negligible increase in the strength of the mortar prisms which were measured in that period, the effect of time on the strength development can be ignored.

2.5.2 *Testing*

The test set-up developed for this research is described in [8]. Figure 2.17 shows the test arrangement which is based on the following principle.

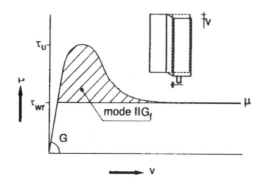
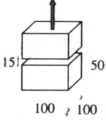

Figure 2.15. Diagrammatical representation of the load shear diagram.

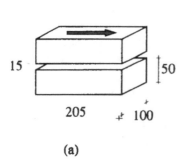

(a)

(b)

Figure 2.16. Shear specimen a) and tension specimen b) (dimensions in mm).

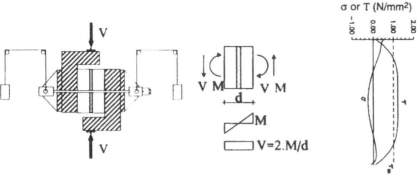

Figure 2.17. Test arrangement and linear elastic stress distribution over the length of the joint.

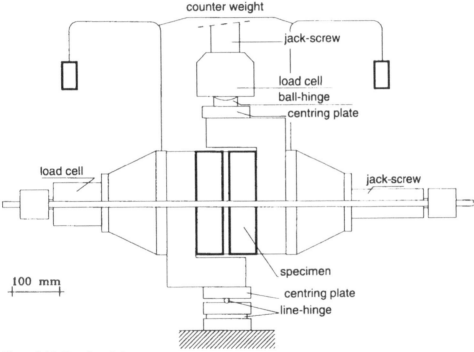

Figure 2.18. Developed shear test set-up.

An externally applied shear force. The line of action runs through the middle of the joint. This force is diverted by stiff steel blocks as a result of which moments and transverse forces are applied to the specimen. This diversion of the force by means of steel blocks is a big difference with many similar shear tests set-ups, in which the diversion takes place in the specimen itself. In case of such tests, normal and shear peak stresses occur, according to Stöckl et al. [20], which are three times as large as the average shear stress. The main feature of the arrangement used is that the shear load causes a rather uniform shear stress distribution in the joint and small normal stresses to the joint. The test set-up that was eventually used is shown in Figure 2.18.

Eight displacements were observed on the specimen: four at the front side and four at the back side. The displacements were registered by means of LVDTs. Figure 2.19 shows the location of the LVDTs on the specimen. Two of the four diagonally placed LVDTs (one on each side) were used to control the increase in deformation. The deformation speed was approximately to 2 mm/hour.

The weight of the L-shaped loading plates and auxiliary equipment to introduce normal stress, was compensated with counterweights. After the zero measurement, the horizontal force was applied perpendicular to the joint and held constant during the test. Subsequently, shear force was applied. Figure 2.20 shows the load shear displacement diagrams of the series S.JG.B. as an example.

The tests do not always develop as indicated in Figure 2.20. When the bond shear strength is relatively high, a diagonal tensile crack may also occur in the units due to the tensile stresses occurring within the units.

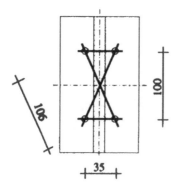

Figure 2.19. Location of the LVDTs on the specimen (dimensions in mm).

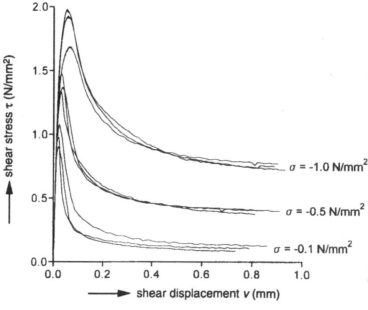

Figure 2.20. Shear stress as a function of the shear displacement for various levels of normal compressive stress in the series S.JG.B.

2.5.3 *Tensile tests with shear tests*

To determine a relationship between the tensile bond strength and the shear bond strength, tension tests were carried out as well, be it this time force-controlled. This was possible since there was no intention to determine the 'post peak' behaviour, contrary to the tension tests in Section 2.4. Besides, these tests enabled us to determine a relationship with the tension tests described in the previous section. Table 2.14 shows the results of the tension tests.

The ST.VE.B. specimens show an average value which is lower than that of T.VE.B (0.22 N/mm^2) from Section 2.4. Possibly the eccentric location of the surface of adhesion plays an important role. This is shown diagrammatically in Figure 2.21. In these tests a hinged attachment was used (Fig. 2.21a) in contrast to Section 2.4 where rotation restrained platens were used (Fig. 2.21b).

The ST.VE.C specimens have more or less the same strength as the T.VE.C specimens.

The ST.JG.B specimens show values that are twice as high and the values of the ST.JG.C specimens are even three times as high as in Section 2.4. The partial fracture in the mortar joint in the ST.JG.C series accounts for the high tensile strength values. The improved bond that is obtained is probably the result of a more optimal moisturization of the units prior to laying. This in contrast to the calcium silicate unit

Table 2.14. Results of force-controlled tension tests on masonry prisms for the comparison of tensile bond strength from previous test series and shear bond strength.

Type of specimen	Tensile bond strength f_{tb} (N/mm^2)	v.c. (%)	Amount of tests
ST.VE.B	0.10	100	5
ST.JG.B	0.62	28	5
ST.CS.B	0.02	72	5
ST.VE.C	0.35	48	5
ST.JG.C	1.43	33	5
ST.CS.C	0.06	22	5

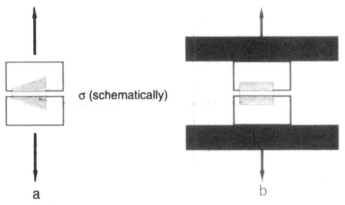

σ (schematically)

a   b

Figure 2.21. The influence of an eccentric surface of adhesion and the method of testing – a) Hinged, b) Fixed – on the stress distribution in the surface of adhesion (diagrammatically).

specimen at which hardly any bond occurred. During the manufacturing the moisture content amounted to 3.7%, to avoid a further decrease in the suction rate. Considering the fact that good tensile bond strength values were found in Section 2.4, on the other hand, and the calcium silicate units there had a much higher moisture content as a result of an immersion for 7.5 minutes, it could be concluded that during the preparation of the calcium silicate unit no attention should have been paid to the suction rate. Experience shows, however, that a moisture content of 5 to 8% in combination with cement lime mortars does not guarantee a good bond strength. It should be noted that in practice the calcium silicate units are processed mainly with a thin layer mortar which results in a better bond, and which is less dependent on the condition of the unit. The next section will show that the shear bond strength especially for calcium silicate unit is considerably higher than the tensile bond strength.

2.5.4 *Shear strength*

This section discusses the shear strength. This is done by means of Mohr-Coulomb's failure criterion. With regard to combined shear and normal stress conditions, this criterion can simply be formulated by:

$$\tau_u = c - \tan(\varphi) \cdot \sigma \tag{2.9}$$

where: t_u = the shear strength, c = the cohesion, φ = the angle of internal friction, and σ = the normal stress to the longitudinal joint (tension is positive).

The cohesion c is also indicated with the shear bond strength τ_o or f_s.

Figure 2.22 shows the maximum shear stress τ_u as function of the normal com-

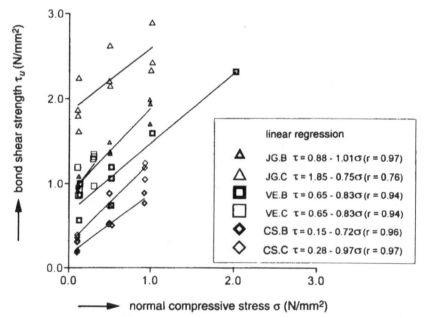

Figure 2.22. Shear strength as function of the normal compressive stress (moisture content calcium silicate unit 3.9% during processing).

pressive stress. The straight lines in the figure have been determined by means of linear regression. The regression analysis gives values for tan (φ) and c.

The behaviour of the specimens with Vijf Eiken units was presumably influenced by a possible compression diagonal which appeared as a result of the presence of the brand mark in the bed face. Besides, the best-fit line of S.VE.C has been determined by means of only 3 tests, so that tan (φ) and c resulting from this line are relatively unreliable. In this series the angle of internal friction is at its largest.

The correlation coefficients r indicated in Figure 2.22, show that the test results do not contradict the assumed linear relation between τ_u and $\sigma(\sigma \leq 0)$. Therefore, Mohr-Coulomb's criterion can be used as failure envelope to describe the shear bond strength.

The angle of internal friction is independent of the materials used in this research, since all the lines in Figure 2.22 run approximately parallel. That does not apply to the cohesion c or shear bond strength f_s with a pre-stress of 0 N/mm^2 (the intersections of the lines with the vertical axis).

Now the tests results show an influence of the mortar quality with specimens of the Vijf Eiken units; in contrast with the results indicated in Section 2.4. Table 2.15 compares the values for c or f_s with the values found before for the tensile bond strength values as indicated in Table 2.14.

Especially in those cases where the tension tests results showed the lowest values, the ratio between the shear bond strength and tensile bond strength is found to be the largest. The values found for the shear bond strength between calcium silicate unit and mortar are considerably higher in comparison with the tensile bond strength values. This is expressed by the high value of the ratio c/f_{tb} for the series KZ.B and KZ.C. Of course higher bond strength values will be obtained where there is a better bond between joint and calcium silicate units. In that case the ratio c/f_{tb} is expected to be smaller.

2.5.5 *Shear stiffness or shear modulus of the mortar joint*

The test results enabled us to determine the shear modulus of the mortar joint.

As with the deformation controlled tensile tests, the measurements have to be corrected with the elastic deformation in the units. The sliding moduli have been determined by means of linear regression from observations up to half the fracture load. This resulted in the values as shown in Table 2.16.

The values used here for the properties of the units are based on compression tests with a low stress level. If we use the values for the elastic moduli of the units from Section 2.4.4, then the resulting values for the sliding moduli of the joints are approximately 10% higher.

Assuming that the elastic modulus of a joint is twice as high as the shear modulus (supposing that the Poisson's ratio $v = 0$), then the values for the mortar joints of the Vijf Eiken units and calcium silicate units do reasonably coincide with the previously found values for the joints in the compression tests. Only the value for KZ.B is significantly lower.

In the case of Joosten masonry this similarity only occurs when the results are compared to the outcome of the compression tests from Series 2 (refer to Section 2.3.5.2).

Table 2.15. Comparison between the tensile bond strength f_{tb} and the shear bond strength f_s (moisture content calcium silicate unit 3.9% during processing).

Type of specimen	Tensile bond strength f_{tb} (N/mm^2)	Type of specimen	Shear bond strength f_s or c (N/mm^2)	c/f_{th} (–)
ST,VE.B	0.10	S.VE.B	0.65	6.5
ST.VE.C	0.35	S.VE.C	0.85	2.5
ST.JG.B	0.62	S.JG.B	0.88	1.4
ST.JG.C	1.43	S.JG.C	1.85	1.3
ST.CS.B	0.02	S.CS.B	0.15	7.5
ST.CS.C	0.06	S.CS.C	0.28	4.7

Table 2.16. Sliding moduli of the mortar joints.

Type of unit	Correction figure E_{unit} (N/mm^2)	v (%)	Type of specimen	G_{mortar} (N/mm^2)	v.c. (%)	Amount of tests
VE	7500	0.10	VE.B	1220	18	7
JG	17500	0.10	VE.C	1330	25	7
CS	15100	0.10	JG.B	2130	8	10
			JG.C	3070	12	11
			CS.B	590	51	7
			CS.C	1300	30	9

2.5.6 *Friction coefficient*

When the friction level was analyzed it became apparent that examination of all the results together would be interesting. In Figure 2.23 the average shear stress from the horizontal part of the curve in Figure 2.15 has been set out against the average normal compressive stress.

Figure 2.23 also shows the average friction coefficients per series with the variation coefficients in brackets and the best straight line through all the test results determined by means of linear regression, including the corresponding correlation coefficient.

From Figure 2.23 it can be concluded that the average friction coefficient amounts to 0.75 and that it is independent of the material that was used. Besides, the cohesion drops back to zero if the shear displacement is large enough. This phenomenon will be discussed in the next section.

2.5.7 *Cohesion-softening and shear crack energy*

It became apparent that when the shear displacement increases, the cohesion does not suddenly, but more or less gradually decreases to zero. This phenomenon shows a great similarity with the results in deformation-controlled tension tests on unit-like materials, in which the tension does not immediately decrease to zero either, once the tensile strength has been reached. The shear fracture energy G_{fII} has been defined in Figure 2.15. In Figure 2.24 the values determined for the fracture energy have been set out against the normal stress that was kept constant during the tests.

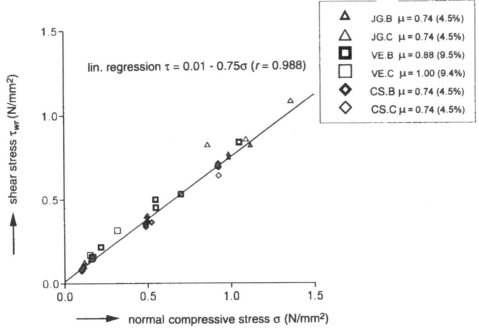

Figure 2.23. Average shear and normal stress in the horizontal end part of the curve of the load displacement diagram.

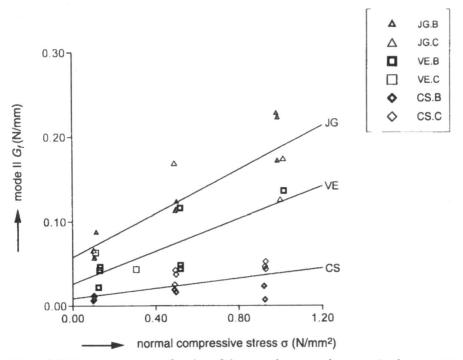

Figure 2.24. Fracture energy as function of the normal compressive stress (moisture content calcium silicate unit 3.9% during handling).

The best straight lines per type of unit resulting from linear regression are also indicated. For each type of unit the fracture energy increases when the normal compressive stress increases. Furthermore the values and the increase in the fracture energy are larger in case of the clay units compared with the calcium silicate units. The type of mortar, as is the case with the shear strength, has no influence here. The fracture energy G_{fII} seems predominantly dependent on the type of unit The linear tendency showed in Figure 2.24 does not imply, however, that a linear relation has been found between σ and G_{fII}. The available data are insufficient to do so. As a rough estimate, however, the linear relations can certainly be used.

To determine the 'post peak' behaviour furthermore, two shear displacements were determined for specimen. These displacements are shown in Figure 2.25. The best-fitting straight line which was determined for the smallest shear displacement v_{lin} by the bold part descending curve had a correlation coefficient $r \geq 0.95$ in most of the tests. The descending curve is therefore almost linear over the observed part. As a result this straight line favourably represents the beginning (the most important part) of the descending curve. This was the reason why the linear fracture energy $G_{fII;lin}$ was determined as well (= shaded area in Fig. 2.25). This area amounted to an average of 60% of G_{fII} with a variation coefficient of 19%.

Figures 2.26 and 2.27 show the displacements v_{lin} and v_{nonlin} as function of the normal compressive stress.

The Figures 2.26 and 2.27 also show that the displacements over which the cohesion reduces to zero, become larger when the values for the normal compressive stress increase. According to the definition $G_{fII;lin} = \frac{1}{2} \cdot v_{lin} \cdot (\tau_u - \tau_{wr})$. When the influence of v_{lin} on $G_{fII;lin}$ is analyzed further, the increase of the softening displacement v_{lin} appears to be the major cause for the increase of $G_{fII;lin}$. This also holds true for G_{fII} and v_{nonlin}.

A formula that can be used to represent the cohesion-softening is Equation (2.10).

$$\tau = c \cdot e^{-\frac{c}{G_{fII}} \cdot v_{nonlin}} \tag{2.10}$$

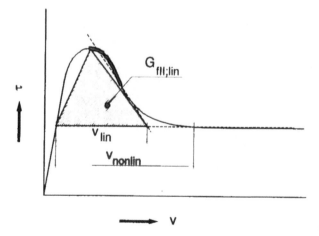

Figure 2.25. Definition of the calculated shear displacements (softening displacements) by means of which the shear stress-shear displacement diagram has been characterized.

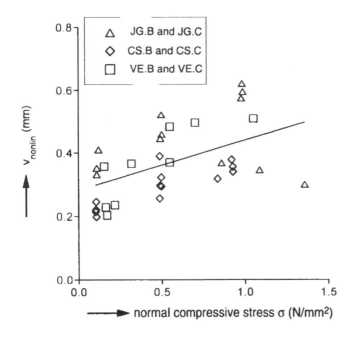

Figure 2.26. Non-linear softening displacements v_{nonlin}.

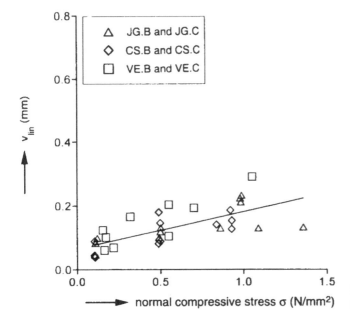

Figure 2.27. Linear softening displacements v_{lin}.

In the Figures 2.28 and 2.29 the test results are compared with the theoretical development of the cohesion-softening according to Equation (2.10). To do so, the value of the cohesion according to Figure 2.22 and the friction coefficient according to Figure 2.23 were used. The picture in the Figures is representative for the tests. From the Figures it can be concluded that Equation (2.10) is not steep enough in the be-

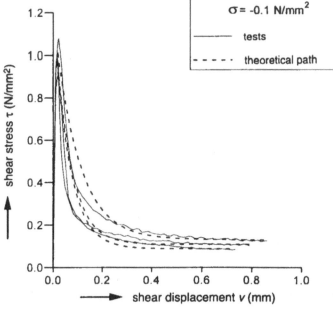

Figure 2.28. Comparison between the tests in series S.JG.B and the theoretical development according to Equation 2.10.

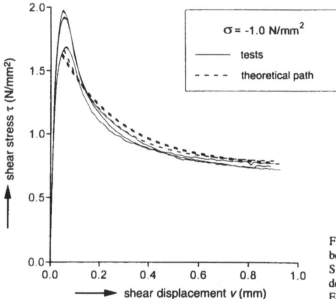

Figure 2.29. Comparison between the tests in series S.JG.B and the theoretical development according to Equation 2.10.

ginning of the descending curve. However, Equation (2.10) appears to be useable as a rough estimate for the cohesion-softening. To describe the first part of the descending curve adequately, a linear approach can also be used very well, as was shown with the determination of v_{lin} and $G_{fII;lin}$.

2.5.8 *Dilatancy*

Another parameter which is also of importance in the numerical modelling of masonry is the dilatancy. This is the phenomenon (as far as the shear tests are concerned) in which, as a result of the relative shear deformation v, also a relative transverse deformation u takes place at right angles to the joint. The angle between u and v is referred to as the angle of dilatancy φ.

The angle of dilatancy ψ was determined from observations in the descending branch after peak load.

Figure 2.30 shows that ψ decreases as function of the normal compressive stress. The dilatancy also appears to depend on the type of unit. Calcium silicate units with their smooth crack surface have the smallest angle. The number of test results is too small to determine whether there is a significant difference between the Vijf Eiken and the Joosten specimen. When the measured transverse displacements are examined roughly, and especially those points at which in some tests the transverse displacement decreases after an initial increase, it can be concluded that the maximum transverse displacement is respectively in the order of 0.1 to 0.15 mm for clay units and 0.05 mm for calcium silicate units.

2.5.9 *CEN tests*

According to Eurocode 6 the shear bond strength has to be determined by means of a shear test on a triplet as described in NNI/CEN norm NEN-EN 1052-3 [24]. A triplet consists of three units and two joints. The test set-up used is shown in Figure 2.31.

NEN-EN 1052-3 does not describe the way in which the load should be applied in the specimens. To lay down the locations where the load should be applied, in a bet-

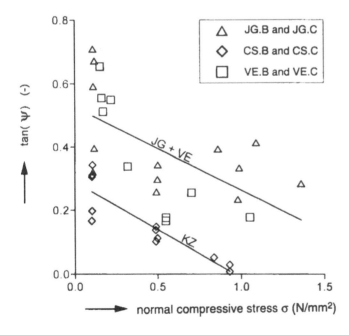

Figure 2.30. Tangent of the angle of dilatancy ψ as function of the normal compressive stress σ.

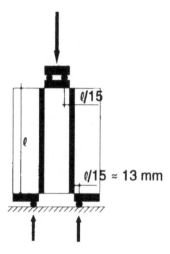

Figure 2.31. Shear test according to NEN-EN 1052-3 with prior conditions according to Riddington et al. [5].

ter way, Riddington [19] proposed to use rollers and to place these at a distance of $l/15$ from the edges (l is the dimension of the unit in stretcher direction).

To come to an international comparison of the test results that were obtained by means of the set-up as described in this report, CEN tests were carried out in fivefold on the Vijf Eiken and Joosten units with mortar C 1:½:4½ as indicated in Figure 2.31.

To make a good comparison between the CEN tests and the deformation-controlled shear tests, tension specimen were manufactured as well by means of which the tensile bond strength was determined. Table 2.17 shows the results.

The absolute differences between both series of tests is probably caused by the higher slump of mortar used for the CEN specimen. If the ratio c/f_{tb} of both test series is compared, the ratios for the CEN shear test are lower. The non uniform stress distribution may be a cause for relatively lower cohesion that was found with the CEN triplet test.

2.6 CONCLUSIONS

2.6.1 *Conclusions with regard to the compression tests*

The compressive strength of masonry is influenced by the units and the mortar that are used. The strength of two-unit-wide specimen consisting of clay unit was about 11% larger than the strength of clay unit specimen that are only one unit wide. This value is nearly equal to that which can be calculated by means of the formula mentioned in the Belgian Norm, NBN 24-301 [22]. In the case of calcium silicate units, the strength of wide and narrow specimen was equal.

The ratio between the elastic moduli of the wide and the narrow specimen was not the same for all types of units and varies from 0.78 to 1.1. Large differences were found between the elastic moduli determined by means of the measured displacements of the loading platens and those determined by means of measurements over

Table 2.17. Shear and tensile bond strength of the CEN-triplet test series and the deformation controlled shear tests series.

Combination of material	CEN-triplet test			Developed shear test		
	c (Mpa)	f_{tb} (Mpa)	c/f_{tb} (–)	c (Mpa)	f_{tb} (Mpa)	c/f_{tb} (–)
VE.C	0.67 (12)*	0.56 (27)	1.2	0.85	0.35	2.4
JG.C	1.38 (10)	1.72 (17)	0.8	1.85	1.43	1.3

*Coefficient of variation in brackets.

112 mm in the middle of the specimens. The elastic modulus of the specimens tended to be largely proportional to the compressive strength.

The elastic moduli of the mortars that were used, have been determined by reducing the deformation of the masonry measured over 112 mm, by the calculated deformation of the units in the measuring area. In addition to this, it is assumed that a unit in a masonry specimen with mortar behaves similar to that in a Bolidt specimen. The range of properties of the units and of the measuring errors come to the fore in the calculated mortar deformation. In the elastic modulus determined that way the deformation of the mortar and the contact surfaces between mortar and unit have been taken into account, whereas a deformation is attributed to the unit which could be expected on the basis of a uniform stress distribution. The elastic moduli calculated this way for the mortars B and C when combined with VE unit, differed by a factor of 2. If these two mortars are used in combination with JG units then the elastic modulus was nearly equal. Contrary to what might be expected, the elastic modulus of the lime rich mortar B 1:2:9 from Series 1 was in case of the Vijf Eiken units larger than that of mortar C 1:½:4½. In Series 1 prefab mortars have been used which were stronger than what might be expected on the basis of their composition. The properties for the mortar are influenced by the units. In the case of units with a small compressive strength the elastic modulus for the mortar is smaller than the elastic modulus for the same mortar when combined with units with a large compressive strength. This difference is presumably caused by the suction rate of the units, which as a rule is smaller in the case of units with a larger compressive strength [17,18]. Observation of the measurements on a single unit on the outside of the specimen, showed that the transducers started to register deformations at different load levels, and that the single unit, especially in the beginning of a test, deformed very little. This is due to the non-uniform stress distribution in the specimen. The first shrinkage cracks appear near the surface during hardening of the specimen which possibly will have to be closed first before compression is applied. The stiffness of the mortaring joint on the outside is also different due to the faster drying out.

When a thin layer of Bolidt was applied between grounded units, the deformations measured on either a single unit, two units together with two joints or the entire specimen, were proportional. With the Bolidt test results a more reliable elastic modulus for units could be established.

The Poisson's ratio v of the unit in the specimen with mortar was not reliable when determined from results taken from the same specimen. What has been established, however, was its tendency to depend on the stress level. By using the elastic moduli of the units from the Bolidt tests in combination with the transverse measurements of the masonry specimen, however, it was possible to determine values for v.

In case of the wide calcium silicate specimens measurements were taken in the width direction on that part of the specimen where a header joint was present. It appeared quite often that the registered deformation did not increase with a rising load, while this did occur with the corresponding narrow specimen. A possible expansion of the unit was obviously compensated by compression of the header joint as a result of which the measured deformation remained smaller than the one measured on the narrow specimen.

Differences between the results of the two compression test series that are discussed, can be accounted for by differences in composition of the mortar and differences in the manufacturing and testing process.

The differences in strength of the mortars used in Series 2 were large. In case of further research, masonry with mortars that have strength values that lie somewhere in between, should surely be included in future tests. Extra attention should then be paid to the composition of the mortars.

The research has shown that the masonry prisms of five units high can be used to determine the masonry compressive strength. Further research into its practicability for structural design is desirable.

2.6.2 *Conclusion with regard to the tension tests*

The elastic modulus of the units has been determined in order to compare specimens among themselves. Since the specimens had a notch, these values cannot be used for other purposes. The moulded units and stuffing press units appear to be 2 to 3 times as stiff in stretcher direction as in thickness direction. The values that were found for the units in stretcher direction show the same tendency as the values from the compression tests in thickness direction. The shape of the specimens played a dominant role here. It is recommended to determine the stiffness of the units in the thickness and stretcher direction by using specimens without a notch.

The stiffness of the joint in tension, based upon the same basic material, varied by a factor of 4 due to the influence of the units between which the mortar of the joint hardens. The mortar B 1:2:9 used in combination with Vijf Eiken units resulted in an elastic modulus of 700 N/mm^2, and an elastic modulus of 2950 N/mm^2 when used in combination with Joosten units.

The tensile strength of the units is of a higher order than tensile bond strength of the joints. The Joosten units had a greater strength in thickness direction than in stretcher direction. The opposite applies to the Vijf Eiken unit. This is the result of the layered composition of the unit owing to the orientation of the clay particles in the unit. The variation of tensile strength values of the unit is limited when compared to the variation of tensile bond strength values. Both the fracture energy for the joints and the tensile strength varied to a large extent. No relationship could be determined between both properties. This coincided with plain concrete for which the crack energy is considered to be a properties of the material, depending on the type of concrete (lightweight concrete, gravel concrete) and which is independent of the strength class of the concrete. Based upon the gross bonding surface, the crack energy of the joint varied between 0.001 and 0.02 N/mm.

Once the net bonding surface has been visually determined, the crack energy of the joint appears to vary between 0.007 and 0.04 N/mm. On average, the crack energy of the joint amounts to 0.02 N/mm, based on the net bonding surface. Based on the gross surface of an continuous wall, the average crack energy is estimated at 0.012 N/mm. Based on the net surface, the variation of tensile bond strength values per specimen decreased considerably. This applied to a lesser extent to the crack energy of the joint. The crack energy of the units has been determined as well. Here the values again are of a higher order than that of the joints and the variation is considerably narrower. Although the crack energy of the joints in the case of calcium silicate units could not be determined, this does not imply that the crack energy is zero. Probably, the crack energy is indeed smaller, since the crack surface area of calcium silicate units was smoother and, therefore, less interlock is expected.

The dimensionless relationship derived by Reinhardt and Hordijk with regard to the stress crack width diagram of concrete can also be used for units and the bonding surface.

2.6.3 *Conclusion with regard to the shear tests*

The results of the tests that were carried out showed that the shear strength of the joint can be described by means of Mohr-Coulomb's criterion when $\sigma \leq 0$. The value for the shear bond strength f_s or cohesion c depended on the type of unit and the type of mortar.

With regard to the combinations of materials that were investigated, the ratio between shear bond strength and tensile bond strength was never less than 1.3. Especially when low tensile bond strength values were found, the ratio increased considerably. This is probably due to the way in which the tensile bond strength has been determined. Patens, connected with hinges to a test rig, were glued onto masonry prisms consisting of two half units and a joint. Especially in the case of an eccentric location of the bonding surface, this will have a negative influence on the strength and that eccentric location occurs more frequently in the case of low strength values.

In the case of calcium silicate units, the bond strength values found in the deformation-controlled tension tests were acceptable and those found in the shear tests were very low. In the shear tests the calcium silicate units were used with a moisture content of 3.4% during fabrication of the specimen to prevent a further decrease of the suction rate. The prescribed regulations with regard to the testing only state that the moisture content of the units should be 5 to 8%. From experience we know, however, that a moisture content of 5 to 8% when using cement lime mortars, does not guarantee a good tensile bond strength. It is recommendable to test the calcium silicate units once more in combination with a dedicated mortar, in order to determine shear properties that are based on a proper bond.

With regard to the angle of internal friction φ a safe estimate is approximately 39°, no matter what type of material is used. Although significantly higher values (maximum 50°) were found in the case of the specimen with Vijf Eiken units, this was probably caused by the brand mark in the unit. A further investigation into the influence of brand marks, recesses, holes etc. in the bed pace on the angle of internal friction is desirable. The friction coefficient is 0.75, independent of the material that was examined. This coincides with an angle of friction of 37°.

The shear crack energy of the joint has been defined and depended on the normal compressive stress. This also applied to the shear displacements over which the cohesion reduces to zero. Lower and upper limits with regard to these displacements have been determined. The shear crack energy varies from 0.009 N/mm in the case of the calcium silicate unit to 0.060 N/mm in the case of the Joosten clay unit when $\sigma = 0$ N/mm^2.

The dilatancy also depended on the compressive normal stress. As the compressive stress increases the angle of dilatancy decreases. This can be accounted for by assuming that with a larger compressive stress the shear planes will polish each other and subsequently become smoother. The Dutch Masonry Standard NEN 6790;1990 TGB Unit Structures states: shear bond strength $f_s = 0.75 f_{tb}$ and the friction coefficient $\mu = 0.2$. These values refer to masonry. When masonry is put in shear various failure modes can occur:

- Tensile bond failure;
- Failure of units;
- Failure of masonry in compression;
- Failure of the mortar joint.

The tests that were carried out indicate that the values mentioned in NEN 6790 are conservative in those cases where the shear strength of the joint is normative. To bring the values in the prescribed regulations to a more realistic level, it is necessary that, apart from shear in the joint, other fracture mechanisms due to shear are examined as well.

Considering the test results found here ($f_s > f_{tb}$, $\mu \geq 0.75$), it is possible to achieve a considerable advantage in comparison to the present regulations in the area where the shear of the joint is normative. It should be noted, however, that in the tests as described in this report only one bed joint with a lengths of ≈ 210 mm has been tested. An investigation with larger masonry specimens, by means of which it is possible to determine the area of validity of the failure mode 'shear of mortar joint', should clarify this.

The shear bond strength values for clay units with mortar 1:½:4½, determined by means of the deformation-controlled shear tests, have been compared with the European shear test NEN-EN 1052-3 (CEN test) which are modified according to Riddington. From a comparison of the ratio f_s / f_{tb} for both the tests, it could be observed that the ratios for the CEN shear test were lower. The non uniform stress distribution may be a cause for the relatively lower cohesion that was found with the CEN triplet test.

2.6.4 *General conclusions*

Simultaneously with the masonry specimens, mortar prisms were made. The deformations determined on these, indicate that the stiffness of these prisms highly exceeds that of the mortar in the masonry specimens. This can be attributed to the totally different conditions with regard to handling and curing of this material as well as the different dimensions. A mortar joint made between sucking units will cure in a different way than a prism made out of the same material in a steel mould. Also the compaction during manufacturing, is completely different in both cases.

In the case of the Vijf Eiken units and the calcium silicate units it appeared during the compression and tension tests that a higher mortar quality does not result in an increased strength and stiffness. Besides, all tests showed that the properties of one and the same mortar, when hardened between different types of unit, can differ considerably.

With the compression test results it was not possible to determine a reliable mathematical relationship between the compressive strength of the units according to NEN 3838 and the mortar strength according to NEN 3835.

Based on the above, the test results on mortar prisms appeared to be only suitable for qualification and quality control of the mortar.

The disadvantage of the method of determination of the values of the elastic moduli calculated for the mortar joints in both the compression and the tension tests is that measuring errors and the variation of unit stiffness were attributed to the joint. Nevertheless, a striking difference between the elastic moduli of the joints in tension and in compression could be observed. It is assumed that this is mainly caused by parts of the area of contact between unit and mortar which are not entirely closed once the mortar has hardened. As a result, it is possible that under pressure at a certain stress level, a larger area of contact between unit and mortar might 'take part' in the transfer of force by which the joint will behave in a more stiffly manner. However, the difference between elastic moduli of the masonry in tension and in compression is considerably smaller owing to the dominant influence of the units on the stiffness of the masonry.

Large variations of bond strength values were found within one series, although a uniform preparation process was aimed for. The level of technology concerning bond is generally too low to consider all factors influencing bond. Subsequently it is impossible to avoid the large dispersions.

The analyses that have been carried out were based on a uniform stress distribution over the entire surface. When the crack energy in the tension tests was determined, for example, it was assumed that the load was applied in the centre line and that the crack would open up uniformly. In reality this is not the case, but in the case of concrete this approach has been applied successfully for some years now.

By modelling finite element models of the test arrangements and using the found non linear behaviour in those models, it is possible to obtain a better insight into the influence of the real distribution of force during a test on the parameters (backward indentification). If necessary, such calculations can serve as a basis to improve material parameters and material models.

CHAPTER 3

Numerical models in DIANA

3.1 GENERAL

The experimental observations from the previous chapter are translated into numerical models in this chapter. The aim was to achieve universal models with regard to the entire course from the linear-elastic stage via the cracking stage up to the failure stage, whereas in many cases the residual behaviour after (partial) failure should be traceable as well. Only then the real 'reserve' of a structure can be determined and is it possible to give a decisive answer about the deformation in the serviceability stage. The necessity to simulate a complex failure behaviour is present, for example, in the case of a vertical structural joint that cracks, which subsequently can transfer a residual force by means of dry friction. A practical relevant example in which prediction of the deformation behaviour is necessary, is the determination of aesthetically acceptable crack widths and crack distances in plastered and non-plastered walls.

This chapter first gives a rough introduction into the non-linear finite element method, where attention is paid to types of elements, treatment of non-linear behaviour and solution procedures. Subsequently, constitutive models for masonry are discussed. This primarily entails discontinuum models for localized crack formation and friction. Besides, attention is paid to unit/block super elements, anisotropic continuous models for masonry as composite, and the scatter in material properties.

3.2 NON-LINEAR FINITE ELEMENT METHOD

3.2.1 *Points of departure and types of elements*

In the finite element method a structure is subdivided into a number of separate 'elements', each of them with certain properties. The method suits masonry extremely well because the elements are already present by nature in the form of units or blocks. The behaviour of the elements is described by means of displacements and forces in the nodes. This (discontinual) is followed by the assembly in which the elements are literally 'knotted together'. Subsequently, the external loads and supports (boundary conditions) are added and a set of equilibrium and compatibility equations is set up for the whole system. The computer is introduced to solve this system of equations, which results in the displacements in the nodes and the support reactions. From the displacements in the nodes the strains and stresses the so-called integration points can be calculated, by which a complete insight into the mechanical behaviour of the structure is obtained. Integration points are needed to achieve numerical integration of the stiffness of the elements. They are, so to speak, sampling points at which the evaluation of material behaviour takes place. Figure 3.1 shows the two types of elements as applied in this research. It concerns a two-dimensional continuum element and a linear line shaped interface element, also referred to as discontinuum element. The continuum element has been used for units and complete masonry units and is based on stress-strain relations. A plane stress situation is assumed here, with three stress components σ_{xx}, σ_{yy}, τ_{xy} and three strain components ε_{xx}, ε_{yy} and γ_{xy}. The collection of components in this report is referred to by means of the stress vector σ and the strain vector ε (superscript ^{T} of transposed):

$$\sigma = [\sigma_{xx}\sigma_{yy}\tau_{xy}]^T \qquad (3.1)$$

$$\varepsilon = [\varepsilon_{xx}\varepsilon_{yy}\varepsilon_{xy}]^T \qquad (3.2)$$

The interface element has been used for crack formation and shear in joints. Here, stress-displacement relations are used instead of stress-strain relations. The stress is either a normal stress σ perpendicular to the interface or a shear stress τ along the interface, with a displacement μ normal to the element (crack width) and the shear displacement v along the element (slip) as corresponding displacements. The collection of these components is called the interface stress vector τ (with τ from traction) and the interface displacement vector μ:

$$\tau = [\sigma\tau]^T \qquad (3.3)$$

$$\mu = [\mu\, v]^T \qquad (3.4)$$

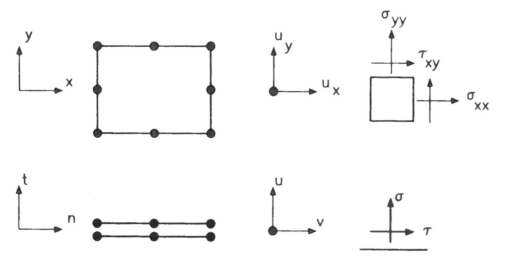

Figure 3.1. Finite elements applied in this research. a) Eight-node plane stress distribution with Gauss integration scheme, b) Six-node interface element with Lobatto integration scheme.

For the continuum elements a Gauss integration diagram was used. With linear-elastic behaviour of the quadratic elements a 2 × 2 Gauss scheme is sufficient, whereas with non-linear behaviour usually a 3 × 3 scheme is necessary. The interface elements have been integrated by means of a Lobatto scheme which for the quadratic elements is identical to a Newton-Cotes or nodal-lumping scheme. Such schemes are necessary to avoid false spurious oscillations [27].

3.2.2 *Solution procedure for non-linear material behaviour*

The central theme lies in the designation of material models to elements. Only in the case of very low stress levels a linear-elastic model can suffice, but in general non-linear models which can simulate crack formation, shear or crushing are needed. The non-linear behaviour of stone-like materials like masonry is characterised by softening (Fig. 3.2). It is known that a material like steel yields at a constant stress. There are plasticity models available which describe this tough behaviour. Another extreme is formed by the class of highly brittle materials, among which glass. Once a little crack appears, such a material will break immediately. The principles of linear-elastic fracture mechanics apply here.

As indicated in Figure 3.2 the behaviour of units, mortar, interfaces, and masonry can be considered as an intermediate form between both extremes. First the stress increases to a maximum to subsequently decrease gradually with increasing deformation. This quasi-brittle behaviour occurs in both tension, compression and shear and is referred to as softening. It is caused by the gradual break down of the weakest links in the heterogeneous material, after which the so developed micro cracks finally link up to form a macro crack or shear plane.

The non-linear relation between load and shear requires an incremental-iterative solution procedure within the finite element method, in which the load is applied step by step (incremental) and equilibrium iterations are carried out with in each load increment. The iteration process is repeated until equilibrium is reached within acceptable limits. This is called a converged solution. In this research the full New

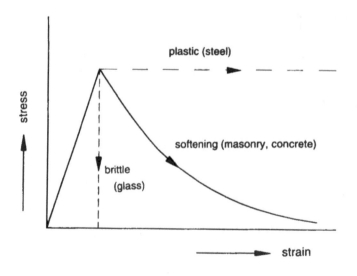

Figure 3.2. Stress-deformation behaviour of materials (diagrammatically). Elastic-plastic, elastic brittle and elastic-softening behaviour.

ton-Raphson iteration procedure is used whereby a new tangent stiffness is set up at the beginning of each iteration. This is shown diagrammatically in Figure 3.3.

The control of the numerical process does not necessarily have to take place by means of load increments. A disadvantage of load increments is that the solution diverges as soon as the global load displacement behaviour approaches a peak, as illustrated in Figure 3.4a. Displacement control offers an alternative. This way the displacement of one or several nodes is incremented so that peaks in the load displacement behaviour can be bypassed (Fig. 3.4b). This method is only suitable for a limited type of problems, in which the force is applied to one node or to a limited number of related nodes.

The most general, third method is arc-length control. This is a kind of combination of load and displacement control by means of which arbitrary paths in the load displacement space can be observed, including reversing curves as illustrated in Figure 3.4c (snap-backs). That these are not academic or fictitious cases will be shown in the practical examples in which indeed such phenomena occur due to the brittle behaviour of some masonry structures. Within DIANA arc-length methods are available in many varieties, of which the method with a selection of degrees of freedom is the most attractive one because it allows an indirect control of the crack opening. For background material, see [28]. It is interesting to note that the possibilities of numerical control run completely parallel to the experimental techniques in this research, in which a deformation controlled method was used as well (refer to Chapter 2).

3.3 JOINT-UNIT DISCONTINUUM MODELS

3.3.1 *Points of departure*

This report primarily aims at a detailed approach in which units and joints are modelled separately. The units are represented by means of continuum elements

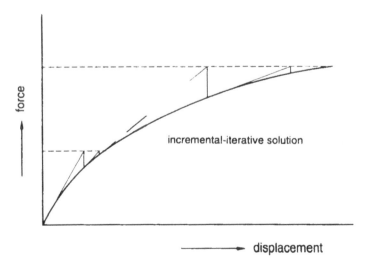

incremental-iterative solution

displacement

Figure 3.3. Incremental-iterative procedure to reach a converged equilibrium solution within the nonlinear finite element method.

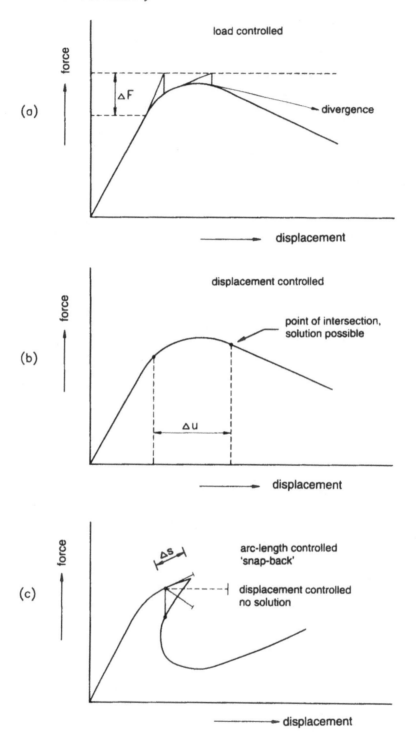

Figure 3.4. Control of the loading process. a) Load controlled. Peaks cannot be by-passed, b) Displacement controlled. Peaks and descending curves can be traced, c) General arc-length method. Also brittle snap-back behaviour can be followed.

(Fig. 3.1a) and the joints by means of discontinuum elements, also called interface elements (Fig. 3.1b). The non-linear behaviour of masonry is than predicted on the basis of the underlying behaviour of the components of units and joint. Literature shows that attempts to such a two-phase approach have already been made before (e.g. Ali & Page [29]), but a significant limitation there was that the softening behaviour that occurred after the peak strength had been reached was not included. It appeared from the concrete mechanics research, among other things, that especially this softening plays a crucial role in the simulation of the crack and fracture behaviour of structures. The models in this report are formulated in such a way that the softening behaviour has been included. This is an innovative element of this research compared to existing masonry literature.

In the joint-unit models in this research it is assumed that the non-linear behaviour is concentrated in the joints whereas the units/blocks have a nearly linear-elastic behaviour. Thus the masonry is looked upon as a set of elastic blocks, connected to each other by non-elastic joints. The joints are modelled, by means of interface elements, as potential cracks or sliding surfaces. This approach agrees with experimental results (refer to Chapter 2), from which it appears that especially the joints and the areas of adhesion between joint and unit contribute to the deformation behaviour of masonry. Furthermore, it appears that with the ratio bond strength/unit strength, as representative of Dutch small-size unit with ordinary mortar, cracks mainly occur along the joints and not, or to a far lesser extent, through the units. This too supports the idea to model joints as the preferential locations for crack and shear formation. In a number of cases in this report the above assumption was deviated from by using a non-linear model not only for the joints but also for the units. This can either be a plasticity model for the non-linear behaviour in compression and/or addition of interface elements in the units in order to allow crack formation right through the units.

3.3.2 *Modelling of units, joints and areas of adhesion*

A second simplification concerns the compound modelling of area of adhesion-mortar-area of adhesion by only one interface element. This is based on the assumption that a distinction between a crack along one or the other area of adhesion is not relevant and is considered too detailed. The situation is illustrated in Figure 3.5. In a fully detailed modelling (Fig. 3.5a) both the unit and the joint would be represented with continuum elements, while interface elements are placed at both areas of adhesion on either side of the joint. Then a crack can run along the upper or lower area of adhesion (the areas of adhesion are usually the weakest links in masonry). An advantage of this modelling is that transverse contraction of the joint is included, which is of importance in the case of collapse due to compression. This has analytically been indicated by Beranek [30] and has numerically been confirmed [31-33]. Figure 3.6 shows a typical result with the essence that the relatively weak joint (a lower E and v than unit) tends to expand under vertical compression but is restrained by the relatively stiff unit, which generates a complex stress field with compression in the joint and lateral tension in the unit. A disadvantage of the complete detailed modelling is that it requires a highly refined mesh, especially where perpendicular and longitudinal joint cross each other.

In the simplified modelling (Fig. 3.5b) only one interface element is used for the

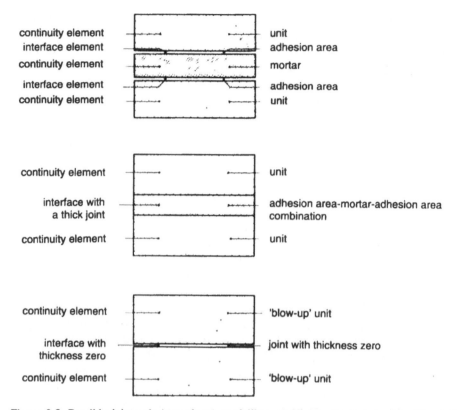

continuity element — unit
interface element — adhesion area
continuity element — mortar
interface element — adhesion area
continuity element — unit

continuity element — unit
interface with a thick joint — adhesion area-mortar-adhesion area combination
continuity element — unit

continuity element — 'blow-up' unit
interface with thickness zero — joint with thickness zero
continuity element — 'blow-up' unit

Figure 3.5. Possible joint-unit (two-phase) modelling. a) Highly detailed model with continuity elements for unit, continuum elements for joint and interface elements for both areas of adhesion, b) Simplified model with continuum elements for unit and an interface element for the combination area of adhesion-mortar-area of adhesion, c) As b), but the units have been 'blown-up' and the interface element has a thickness zero.

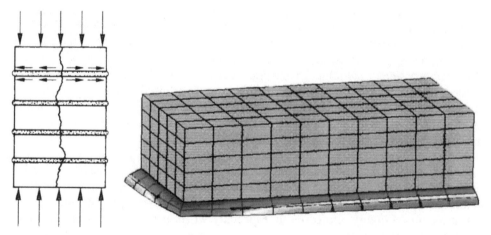

Figure 3.6. Illustration of interaction behaviour between unit and joint, caused by differences in elastic properties [31, 32].

joint, including both areas of adhesion, so that considerably fewer elements are necessary, which reduces the calculation time. A disadvantage here is that the effect of transverse contraction of the relatively weak joint on the relatively stiff unit is neglected in compression. This can be justified for problems in which not the behaviour under compression but the behaviour under tension, bending or shear is normative. This was shown in a pilot study with comparable calculations for piers in bending [34]. What complicates the simplified modelling, is the need for a correction with regard to the thickness of interface element.

Usually the scheme according to Figure 3.5b would be applied, in which the compound interface element indicating area of adhesion-joint-area of adhesion has a thickness equal to the thickness of the joint. This scheme, however, does not meet the requirement of moments equilibrium, which could lead to an error of 10 to 20%, dependent on the real ratio of the thickness of the joint and the thickness of the unit [34]. This is caused by the fact that the thickness has not been taken into account in the formulation of a interface elements. Interface elements only function properly when the thickness is zero. As a remedy the scheme of Figure 3.5c has been applied in this study, in which the units were 'blown-up' in order to give the compound interface element indicating area of adhesion-mortar-area of adhesion, a thickness of zero.

3.3.3 *Linear-elastic behaviour*

At the beginning of a non-linear calculation the stress levels are relatively low, and, therefore, a linear-elastic model is sufficient. The constitutive behaviour of the unit is described by stress-strain relations for continuum elements. The following applies to the linear-elastic stage:

$$\sigma = D^e \varepsilon \tag{3.5}$$

with σ and ε according to Equations (2.9) and (2.10), ^{e} from elastic, and D^e according to Hooke's law:

$$De = \frac{E}{1 - v2} \begin{bmatrix} 1 & v & 0 \\ v & 1 & 0 \\ 0 & \times & \frac{1-v}{2} \end{bmatrix} \tag{3.6}$$

where E is the elastic modulus and v is the Poison's ratio.

The constitutive behaviour of area of adhesion-joint-area of adhesion is described by stress displacement relations for the interface elements. The following applies to the linear-elastic stage:

$$t = K^e u \tag{3.7}$$

with t and u according to Equations (3.1) and (3.2). The stiffness K^e in the linear-elastic stage is assumed to be:

$$K^e = \begin{bmatrix} k_n & 0 \\ 0 & k_t \end{bmatrix} \tag{3.8}$$

with the normal stiffness k_n perpendicular to the interface and the shear stiffness k_t along the boundary layer.

These stiffness values for boundary layer elements have a dimension N/mm³, contrary to the elastic moduli for continuum elements which have a dimension of N/mm².

The constants in Equations (3.4) and (3.6) are determined on the basis of the experimentally measured elastic moduli for units, E_{unit}, and for joints including the areas of adhesion, E_{joint}. The experimental values have been reported in Chapter 2.

Two matters of interpretation play a role in that. Firstly, the tests emphatically show differences for tension and compression. Anticipating the expected stress situation (primarily compression, tension or combinations) a choice has been made in the examples for the elastic modulus in tension, the elastic modulus in compression or an average of the two. A possible alternative could be to model the kink in the origin directly, but the effect of this has so far not been investigated.

A second problem concerns the fact that the experimentally determined elastic moduli are valid for the actual dimensions of joint and unit. A translation is needed to finish up with the 'blown-up' unit with 'zero thickness' joint as shown in Figure 3.5c. This procedure is sketched in Figure 3.7.

It is required that the total lengthening Δl_{tot} across unit and joint is equal in both cases. The following applies to the actual situation (Fig. 3.7a):

$$\Delta l_{tot} = \sigma \frac{h_{unit}}{E_{unit}} + \sigma \frac{h_{joint}}{E_{joint}} \tag{3.9}$$

with the thickness of the brick, h_{unit}, the elastic modulus of the unit, E_{unit}, the thickness of the joint, h_{joint} and the elastic modulus of the joint including both areas of adhesion, E_{joint}. The following applies to the model situation (Fig. 3.7b), indicated by an apostrophe:

$$\Delta l'_{tot} = \sigma \frac{h'_{unit}}{E'_{unit}} + \sigma \frac{1}{k_n} \tag{3.10}$$

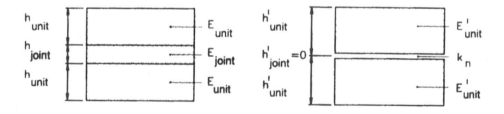

$$h'_{unit} = h_{unit} + h_{joint} \left.\begin{array}{c} \\ \\ \end{array}\right\} \quad k_n = \frac{E_{unit}\, E_{joint}}{h_{joint}(\,E_{unit}\, E_{joint})}$$
$$E'_{unit} = E_{unit}$$

Figure 3.7. Scheme to determine the stiffness k_n according to Equation (3.6). a) Real situation, b) Modelling with 'blown-up' unit and zero thickness joint.

with the thickness of the blown-up unit, h'_{unit}, the elastic modulus of the blown-up unit, E'_{unit}, and the normal stiffness of the boundary layer element with zero thickness, k_n. Based on the requirement

$$\Delta l_{tot} = \Delta l'_{tot} \tag{3.11}$$

the necessary stiffness k_n can be determined. In order to do so, we assume that the blown-up unit has the same elastic modulus as the real unit:

$$E'_{unit} = E_{unit} \tag{3.12}$$

A possible alternative could be to reduce E'_{unit} in relation to E_{unit} but the resulting influence is considered to be small. Furthermore, because of the interface element with a thickness of zero, the following applies:

$$h'_{unit} = h_{unit} + h_{joint} \tag{3.13}$$

The combination of Equation (3.7) up to and including Equation (3.12) leads to the following stiffness of the zero thickness interface element in Figure 3.7b:

$$k_n = \frac{E_{unit} E_{joint}}{h_{joint}(E_{unit} - E_{joint})} \tag{3.14}$$

A similar line of reasoning can be applied to the shear stiffness k_t of the interface element, which leads to:

$$k_t = \frac{G_{unit} G_{joint}}{h_{joint}(G_{unit} - G_{joint})} \tag{3.15}$$

with the shear modulus G_{unit} of the unit, and the shear modulus G_{joint} of the joint including both areas of adhesion. With regard to the unit the regular relationship between shear modulus and elastic modulus applies:

$$G_{unit} = \frac{E_{unit}}{2(1 + v_{unit})} \tag{3.16}$$

with v_{unit} being Poisson's ratio of the unit, it is interesting that this relationship more or less applies to the joint including both areas of adhesion as well [11]:

$$G_{joint} = \frac{E_{joint}}{2(1 + v_{joint})} \tag{3.17}$$

despite the fact that G_{joint} and E_{joint} are compound quantities for mortar plus both the areas of adhesion.

In case of comparatively thin joints the translation from Figure 3.7 can be omitted. This is the case, for example, with calcium silicate products where large elements of 300×600 mm are glued together with by thin high performance joints with a thickness of 3 mm. In such cases it is possible to give the interface element the actual thickness of the joint, instead of a zero thickness. The error made with regard to the equilibrium can be neglected then. The following applies directly to the stiffness:

$$k_n = E_{joint} \, / \, h_{joint} \qquad\qquad (3.18)$$

and

$$k_t = G_{joint} \, / \, h_{joint} \qquad\qquad (3.19)$$

3.3.4 *Discrete crack formation*

The linear-elastic relations from the previous section should be limited in the form of models for crack formation and for friction. Crack formation occurs as soon as a tensile stress or a combination of tensile and shear stress exceeds a tension cut-off criterion. Friction/slip occurs as soon as the combination of compressive and shear stresses exceeds a failure envelope. The interface model should represent both these phenomena. This section deals with crack formation, the next section will deal with friction and subsequently the combination of both models.

Since interface elements (discontinua) are used, the terminology discrete cracking is adopted, as a counterpart of the so-called smeared cracking for continua. The initiation of a discrete crack has in this study been modelled by means of a tension cut-off criterion consisting of a straight vertical line in the σ, τ (Fig. 3.8). This criterion involves only one parameter: the uni-axial tensile strength f_t which for joints is equal to the bond tensile strength (assuming that the area of adhesion is the weakest link) and for units is equal to the uni-axial unit tensile strength. The value of f_t can be derived directly from the tension tests from Section 2.4.

After the formation of the crack, the tensile stress does not immediately drop to zero but gradually. This tension-softening behaviour prior to crack propagation has already been introduced in Chapter 2 and Section 3.2.2. The tension tests from Section 2.4 indicate a softening curve which in the beginning rapidly descends and

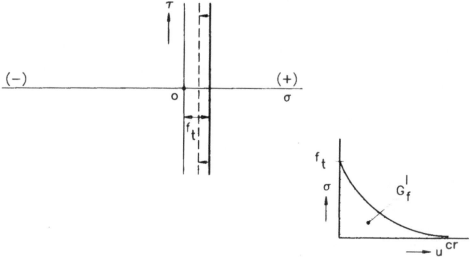

Figure 3.8. Model for tensile crack formation in interface elements. Tension cut-off for crack initiation. Tension softening for crack propagation.

subsequently shows a long 'tail'. This feature has been transformed into an exponential softening-function to indicate the relationship between the tension stress σ and the crack opening u:

$$\sigma = f_t e^{-\frac{f_t}{G_f^I} u} \tag{3.20}$$

G_f^I is the Mode-I fracture energy, defined as the amount of energy required to create one unit area of a tensile crack (dimension $[J/m^2]$ or $[N/m]$). Mode-I is a term originating from fracture mechanics referring to a crack in which only opening displacements occur between the crack faces but no shear displacements. Integration of Equation (3.20) results in G_f^I, which implies that the Mode-I fracture energy corresponds with the area under the tension softening curve, as shown in Figure 3.8. The value of G_f^I has been determined by means of the tension tests from Section 2.4. The softening-function from Equation (3.20) approaches the experimental results as precisely as the function represented in Section 2.4 (the differences are only marginal), but from a numerical point of view it is easier. For this reason it is preferred.

Once the tension cut-off has been reached, not only the relation between the crack normal stress and crack width u is necessary, but also an assumption concerning the shear stress has to be made. Initially, a shear stress was used which immediately drops to zero once the crack appears. This method was rather successful with relatively small shear stress values, that is to say in situations in which the tension cut-off was intersected at or near the horizontal axis of Figure 3.8. In the case of larger shear stresses at the onset of cracking, the sudden drop caused convergence problems. For this reason a model has been developed in which the tension-softening relation was not taken into account in the direction perpendicular to the interface, as usual [35, 27], but in the direction of the stress resultant $\sqrt{\sigma^2 + \tau^2}$ at the onset of cracking [36]. At the same the onset of cracking, the linear stiffness relation from Equations (3.7) and (3.8) is replaced by

$$\Delta t = T^T K^{cr} T \Delta u \tag{3.21}$$

Here T stands for the transformation matrix from the n, t interface directions to the inclined softening-direction. Δ is an indication that an incremental relationship is involved, as needed in the non-linear finite element method. K^{cr} is the stiffness matrix in the inclined softening-direction, according to

$$K^{cr} = \begin{bmatrix} k_n^{cr} & 0 \\ 0 & 0 \end{bmatrix} \tag{3.22}$$

with k_n^{cr} being the negative slope of the softening diagram (Fig. 3.8 and Eq. 3.16). The shear stiffness k_t^{cr} along the direction perpendicular to the softening-direction is supposed to be zero. This assumption appeared to give the best results. An assumption was necessary because experimental data for tension/shear-combinations are not available.

Sometimes, in discrete crack models a decomposition of the interface displacements u and v is applied in an elastic part and a cracked part [27]. When the elastic deformation in the interface-element is small in comparison to the crack deformation this decomposition can be omitted. In this research the latter option was chosen, also called a 'total' formulation. An advantage of the total formulation is that an internal iteration-loop, needed with regard to the decomposed formulation, can be omitted. This simplifies the calculation considerably. A disadvantage of the total formulation is the presence of a discontinuity in the model at the onset of crack formation; the elastic deformation that is present is suddenly transformed into crack deformation, which is accompanied by a drop of the tensile stress. The magnitude of the discontinuity is dependent on the ratio between k_n in Equation (3.8) and k_n^{cr} in Equation (3.22). In case of a relatively high k_n the drop in stress remains limited. As a high elastic dummy stiffness is usually used with interface elements (the elastic deformation is often modelled in the continuum elements on either side of the interface element) no difficulties are expected.

3.3.5 *Coulomb friction*

With regard to shear stresses and combinations of shear stresses and compressive stresses along the interface, Coulomb's friction criterion has been applied. Figure 3.9 shows the criterion as failure envelope in the σ, τ space. It is defined by two parameters: the cohesion c_u, which is the maximum shear stress at zero normal stress, and the angle of internal friction ϕ, also expressed as coefficient of friction $\mu = \tan \phi$.

The criterion can be expressed by the following formula:

$$|\tau| \le c_u - \sigma \tan \phi \tag{3.23}$$

in which compressive stresses are negative. As soon as the combination of compressive stress σ and shear stress τ has reached the envelope, friction/slip occurs. During this process the cohesion c will not keep its maximum value c_u, but will gradually decrease with increasing slip. This cohesion-softening complies with a translation of the envelope to the left, as shown in Figure 3.9.

Cohesion, angle of internal friction and cohesion-softening have been determined experimentally by means of shear tests (refer to Section 2.5). As is the case in the

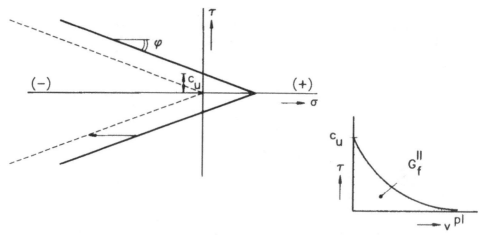

Figure 3.9. Model for Coulomb friction in interface elements. Failure envelope and cohesion-softening for shear crack/sliding surface.

tension tests, the shear tests indicate that there is an exponential softening-relation. The relation between the cohesion c and the slip v^{pl} (superscript pl stands for plastic) has been modelled according to:

$$c = c_u e^{-\frac{c_u}{G_f^{II}} vpl}$$ (3.24)

G_f^{II} is the Mode-II fracture energy, defined as the amount of energy that is needed to create one unit area of a shear crack. Mode-II is a term originating from fracture mechanics referring to a crack in which only a parallel displacement occurs between the crack faces but no crack width. Integration of Equation (3.24) results in G_f^{II}, which implies that the Mode-II fracture energy corresponds with the area under the cohesion-softening curve, as indicated in Figure 3.9. The value of G_f^{II} and the exponential shape of the cohesion-softening curve have been determined by means of the shear tests from Section 2.5.

The friction model has been formulated in the context of plasticity. This implies a decomposition of the interface displacements u and v in an elastic and a plastic part. Here the concepts of yield function and flow rule are used.

Equation (3.23) is used as yield function, i.e. the failure envelope from Figure 3.9 is the yield criterion, that gradually reduces as a consequence of softening with an increasing plastic slip v^{pl}. The flow rule determines the increase of the plastic deformation during slip. This requires a fourth parameter in addition to c, ϕ and G_f^{II}: the angle of dilatancy ψ, which indicates the amount of normal displacement under shear (dilatancy, sometimes incorrectly mixed up in the unit laying community with dilatation, which also indicates expansion in a joint but for a different reason.) The definition is as follows:

$$\tan \psi = \frac{\Delta u^{pl}}{\Delta v^{pl}}$$ (3.25)

The meaning has been illustrated in Figure 3.10. Dilatancy is highly relevant in case of confinement. The material wants to expand but is obstructed from doing so, which causes a wedging effect and high compression stress values. Some examples will be given in this report. When $\psi = \phi$ this is called associated plasticity and when $\psi < \phi$ this is called non-associated plasticity. The shear tests from Section 2.4 result in an

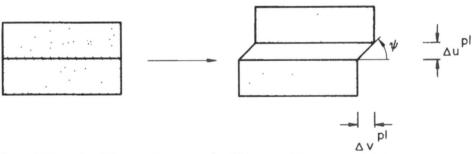

Figure 3.10. Angle of dilatancy ψ: amount of uplift in case of shear across a joint.

average tan $\phi = 0.75$ and tan $\psi = 0.2$, dependent on the roughness of the surface of joint and unit. For this reason non-associated plasticity has been used in combination with special solution procedures for the resulting non-symmetrical systems, as available within DIANA.

For the elaboration of the elasto-plastic tangent stiffness matrix K^{ep} refer to [28, 37]. The apex in the yield criterion requires a special treatment. For a consistent formulation refer to [37]. For most analyses in this report a simplified procedure has been applied. As soon as the stress combination reaches the apex during cohesion-softening, both σ, τ as well as the stiffness have immediately been set to zero. This is based on the assumption that the resulting damage after the apex has been reached is that large, that there will hardly be any residual stress left. The assumption of a sudden stress reduction appeared to be acceptable in most cases because the apex is only reached at a late stage, after a certain amount of cohesion-softening. The apex will not be reached directly under uni-axial tension or tension that is nearly uni-axial, because for such stress combinations the tension cut-off criterion is added, as described in the next section. Only stress points that drift to the apex during cohesion-softening are liable to the sudden reduction of stress.

3.3.6 *Combination cracking and Coulomb friction*

The crack model primarily applies to tension and the friction model to shear and compression/shear. A natural choice for a general model is then the combination of both, as illustrated in Figure 3.11. This combination has been applied in most of the analyses in this project.

It is assumed that an integration point of the interface elements is either cracked or plastic (sheared), but not simultaneously or successively cracked as well as plastic. A shift from crack formation into plasticity or the other way around is not allowed. As

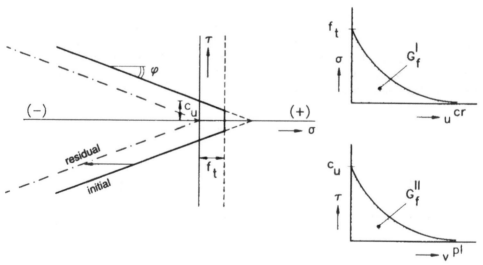

Figure 3.11. Combination Coulomb friction with tension cut-off. Applied in most of the calculations in this report.

soon as the damaging process starts with crack formation and tension-softening, no subsequent check is made to see whether Coulomb friction is also taking place, and, vice versa, as soon as the damaging process starts with Coulomb friction and cohesion-softening, no subsequent check with regard to the cause of a tensile crack takes place. This simplification appeared to be acceptable for the calculations in this report. The acceptance implies that despite the 'cutting' of the Coulomb friction criterion, it is still possible to reach the apex, as indicated in the previous section.

3.3.7 *Alternative formulations*

The combined crack formation/friction model from Figure 3.11 should be looked upon as a compromise which, after verification and evaluation (Chapter 4), was considered good enough to carry out the practice-oriented case studies. Pursuing further refinements and improvements had no priority at that moment.

This does not mean, however, that alternative formulations were not tested in the start-up phase. One of them concerned the assumption of a parabolic failure criterion, fully formulated in the context of plasticity (Fig. 3.12a). The parabola has the following formula:

$$\sigma = -\frac{1}{4f_{tu} \tan^2 \phi} \tau^2 + f_t \qquad (3.26)$$

where f_{tu} is the ultimate value of the tensile strength (an arbitrary tensile strength, not to be confused with the same symbol for the unit tensile strength from Section 2.4), f_t is the residual value from the tensile strength during softening and $\tan \phi$ is the gradient of the tangent line when $f_t = f_{tu}$ and $\sigma = 0$. Softening has been included by the gradual decrease of f_t in Equation (3.26) from its ultimate value f_{tu} to zero, via an exponential softening function in analogy with Equations (3.20) and (3.24). The softening formulation then coincides with a translation of the parabola to the left (Fig. 3.12a), in which f_{tu} corresponds with the top of the initial parabola and f_t corresponds with the top of the shifted parabola. Either the plastic shear/slip v^{pl} can be used as softening-parameter, in analogy with Equation (3.24) for Coulomb friction, or the plastic normal displacement at u^{pl}, playing the same role as the crack width u^{cr} in Equation (3.20) for discrete cracks, or combinations of both.

The parabolic criterion, like Coulomb friction, is a two-parameter model. Above, the parabola has been formulated in f_t (initially f_{tu}) and $\tan \phi$. If we define the cohesion c_u as the value of τ when $\sigma = 0$ for the initial parabola, the following applies:

$$c_u = 2f_{tu} \tan \phi \qquad (3.27)$$

From this it appears that the parabolic criterion can also be formulated with f_t (initially f_{tu}) and c_u as follows:

$$\sigma = -\frac{f_{tu}}{c_u^2} \tau^2 + f_t \qquad (3.28)$$

The relationship between c_u and f_{tu} Equation (3.27) coincides favourably with the material testing from Chapter 2 which indicates that $c_u = 1.5 f_{tu} \tan \phi$ up to $2 f_{tu} \tan \phi$.

The parabolic plasticity model was implemented and tested in a number of vari-

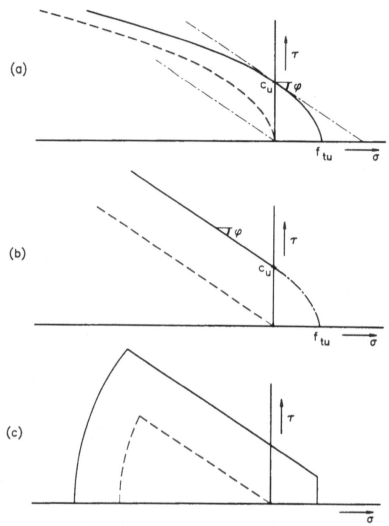

Figure 3.12. Some alternative models for interface behaviour. a) Parabolic plasticity criterion for tension/shear and compression/shear, b) Coulomb friction for compression/shear and parabolic tension cut-off for tension/shear [36], c) Composed multi-surface plasticity formulation with tension cap and compression cap [37].

ants. A second-order tangent-stiffness matrix was used which takes into account the curvature of the yield criterion. An application concerning a pier-wall connection is discussed in [38]. An advantage of the model is the use of plasticity with regard to both the tension/shear and the compression/shear area. In the case of the combined crack formation/friction model from Figure 3.11 the friction part was formulated via plasticity, but the tension part via discrete cracking. Restrictions had to be applied there as soon as friction was followed into crack formation or vice versa, as described in Section 3.3.5. This is not the case with the parabolic plasticity criterion.

A disadvantage, on the other hand, is that the shape of the parabola does not coincide with the micro-shear tests from Chapter 2. With regard to the tension/shear area a suitable interpolation was obtained between the pure tension tests and pure shear by means of Equation (3.27), but with regard to the compression/shear area the curvature of the parabola is too large, especially in the residual stage after softening. In case of translation of the parabola to the left, the real residual strength is overestimated. This is shown diagrammatically in Figure 3.13, in which the dotted parabola considerably deviates from the available measuring points in the residual stage. In the application in [38] the residual strength of the structure for this reason appeared to be predicted wrongly. The problem can be solved if during softening the parabola is not only translated, but is also given a different shape, for example, by making the curve more slender. Another possibility is the application of a third-degree curve or a hyperbola. Such attempts have not been made so far.

A second alternative model has been illustrated in Figure 3.12b. Here the parabola has only been applied with regard to the tension/shear area as tension cut-off in a discrete crack formulation, while for the compression/shear area Coulomb friction was used. This is a logical answer to the restriction of the complete parabola as mentioned above. The fit of the micro-shear tests according to Figure 3.13 is optimal now, while discontinuities are avoided. When $\sigma = 0$, a smooth transition with tan ϕ as a gradient is obtained. Examples with this model have been reported in [36, 39]. What remained a disadvantage was the fact that restrictions had to be made once again with regard to the change from plasticity to crack formation in an integration point and vice versa. When applied in practice-oriented case studies, the advantages in comparison to the model from Figure 3.11 appeared to be only marginal.

Recently, a third alternative was worked out (Lourenco, Rots and Blaauwendraad [37]). This concerns a complete plasticity definition with a composite yield criterion for both tension, compression and shear, Figure 4.12c. The criterion can be looked upon as Coulomb friction with a cut-off on both the tension side and the compression side. In the first instance, a straight tension cut-off and an elliptic compression cap

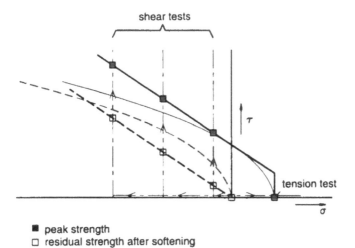

■ peak strength
□ residual strength after softening

Figure 3.13. Available experimental data for peak strength values and residual strength values from micro-tension and micro-shear tests from Chapter 2. Possible approaches (diagrammatically).

were used, although combinations of other shapes and angles are possible as well. This is because the corner points have been treated consistently according to multi-surface plasticity. With regard to the three domains various hardening/softening parameters and laws can be used, whereby also coupling of, for example, Mode-I tension and Mode-II shear damages can be taken into account. The compression cap was added to restrict the open Coulomb friction criterion. The physical background being the crushing of the joint or the damage of the masonry composite under high compressive stresses. The acceptance of a compression restriction apart from a restriction of tension and shear favourably matches Mann's well-known masonry research report [40].

Although the development is still in the start-up phase, it can already be concluded that the plasticity approach with a composite yield criterion looks promising. It favourably matches the experimental observations and there is an optimal numerical stability. A limitation, however, is the fact that the behaviour upon unloading is always elastic. It is more difficult to describe the closing and reopening of tension cracks, whereas this is quite simple in the case of a discrete crack formulation. There, the direction of the crack is memorized and closing/reopening with regard to that direction can be described by means of a secant approach [27]. To what extent this limitation of plasticity formulations plays a role, naturally depends on the specific application.

Failure due to compression was not decisive in the practice-oriented case studies that have been studied in this research projects pier-wall connections (crack behaviour of walls under restrained shrinkage, diaphragm walls). The fact that the compression cap was not included in those studies was justified for that reason. The restriction according to Figure 3.11 with regard to shear and tension was sufficient.

3.4 SUPER ELEMENTS

One feature of all the models described above is that the softening is concentrated in interface-elements, whereas the behaviour of the units or masonry parts on both sides of the interface elements remains linear-elastic.

The assumption of linear-elastic behaviour of the surrounding continuum enables time saving with regard to computer time via substructuring. This technique replaces a combined series of linear-elastic elements by a so-called super element. Figure 3.14 shows that the internal nodes of the substructure are eliminated or condensed, so that solely the boundary nodes remain. The boundary nodes should be maintained because they form the connection with the non-linear interface elements. The system of equations is drastically reduced this way, which reduces the calculation time. At least as important is that the internal force vector no longer needs to be formed via all the individual integration points of the original elements, but directly by means of the manipulation $F_{substr} = K_{substr;linear-substr}$ for the linear-elastic super element. This saves many computer-intensive manipulations at integration point level.

The technique of super elements fits in extremely well with masonry. This was already recognized by Ali & Page [29]. The large elements, blocks or units are inherently present and behave almost linear-elastically. Apart from cases in which individual blocks or units are examined, super elements can also be of use when

unit/block super element

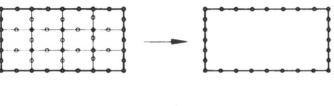

Figure 3.14. Substruc-
turing with unit/block
super elements to speed
up the calculation pro-
cedure.

• boundary nodes
○ internal nodes eliminated

complete masonry parts are modelled as composite. The analysis of crack formation in long walls [41] is an example of that. The location where the crack will develop is known beforehand. Only a single row of interface-elements is needed and the large masonry parts on both sides of the crack can be replaced by two super elements.

Partly within the framework of this CUR-project, the substructuring technique was implemented in DIANA. This was carried out and reported by Nauta [42]. The extent of calculation time saving is of course dependent on the ratio between the number of eliminated nodes and boundary nodes. In the case of a shear wall with a relatively course element mesh a factor of 3 was reached. That example has been included in the test suite of DIANA. In the case of very fine subdivisions in the super element the saving of time can be enormous. With simulations of crack propagation in complete walls acceleration by a factor of 5 to 10 is no exception.

3.5 ANISOTROPIC CONTINUUM MODELS FOR MASONRY AS A COMPOSITE

3.5.1 *Points of departure*

The models discussed so far can be ranked under the common denominator of dis-continuum models with regard to crack formation and slip in interface elements. Since the damage in unreinforced masonry – which this study is restricted to – is a localised phenomenon by definition, the discontinuum approach is usually adequate, especially when the location of crack and slip lines can be determined beforehand by means of engineering judgement.

There remains one category of problems which requires a continuum approach for the masonry composite. Figure 3.15 gives an example of a complete facade where a detailed simulation of the crack propagation along individual units and joints is nei-ther possible nor desirable. In such cases the objective is to describe the effect of joints, units and stacked structures in a smeared sense. This leads to a composite model with anisotropic properties. It should be obvious that the method is especially suitable when the dimensions of the finite element are large enough in comparison to the unit/block dimensions, as illustrated in Figure 3.16. Applications mainly occur with global analyses and in situations in which the location of the crack(s) is/are not

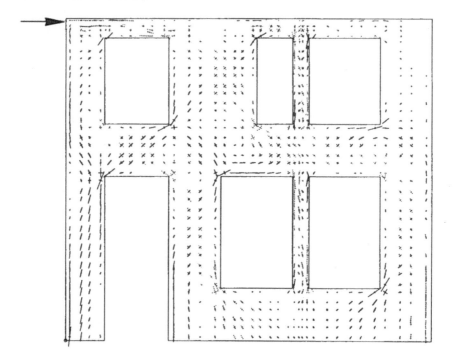

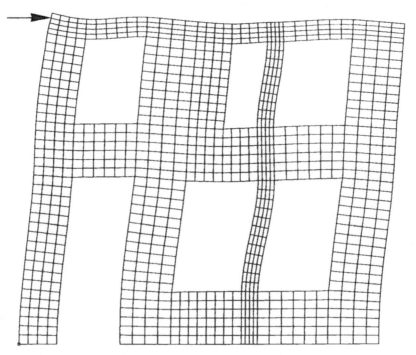

Figure 3.15. Example of global continuum modelling of masonry as a composite. a) Principal stresses in uncracked stage (full line: tension, dotted line: compression), b) Deformations in un-cracked stage.

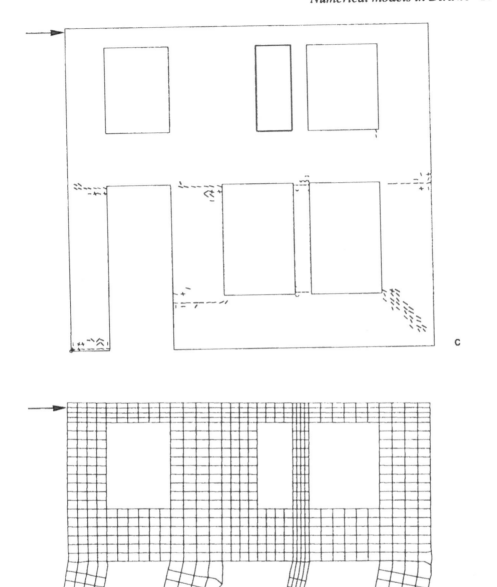

Figure 3.15. Continued. c) Smeared cracks in areas where the tensile strength has been reached, d) Deformations in cracked stage. Slanting mechanism is visible.

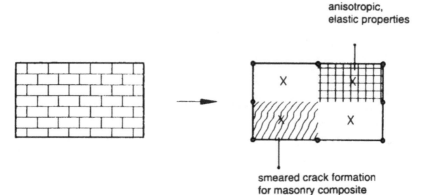

Figure 3.16. Anisotropic properties and smeared crack formation with regard to the masonry composite.

known beforehand. The facade with openings in Figure 3.15 being an example of that.

Figure 3.15a shows the distribution of principal stresses in the uncracked (linear-elastic) stage and Figure 3.15b shows the accompanying deformations. We can observe maximum principal tension stresses on the corners of the openings and in the tension zones of the slender piers in bending. In these areas cracks will appear when the load increases. In such calculations the cracks are not modelled between the elements but inside the elements. Figure 3.15c shows the smeared cracks in the integration points of the elements. From the accompanying deformations (Fig. 3.15d) it can be concluded that a slanting mechanism had been predicted for this example.

3.5.2 *Orthotropic-elastic behaviour*

The elastic modulus of masonry in the direction perpendicular to the longitudinal joints $(E_\perp)$ is lower than the elastic modulus parallel to the longitudinal joints $(E_{//})$. According to the Netherlands Codes [43] this difference can even amount to a factor of 2, dependent on the type of masonry. With regard to the initial stiffness of the composite an orthotropic linear-elastic model should be used. The isotropic relation from Equation (3.6) is replaced by the following orthotropic relation:

$$
\begin{bmatrix} \sigma_{xx} \\ \sigma_{yy} \\ \sigma_{xy} \end{bmatrix} =
\begin{bmatrix}
\dfrac{E_{//}}{1 - v^2 E_\perp / E_{//}} & \dfrac{v E_\perp}{1 - v^2 E_\perp / E_{//}} & 0 \\[3ex]
\dfrac{v E_\perp}{1 - v^2 E_\perp / E_{//}} & \dfrac{E_\perp}{1 - v^2 E_\perp / E_{//}} & 0 \\[3ex]
0 & 0 & G
\end{bmatrix}
\begin{bmatrix} \varepsilon_{xx} \\ \varepsilon_{yy} \\ \gamma_{xy} \end{bmatrix}
\tag{3.29}
$$

where x is the direction parallel to and y is the direction perpendicular to the longitudinal joints. The model has four parameters: $E_{//}$, $E_{\perp}$, Poisson's ratio v and the shear modulus G.

3.5.3 *Smeared cracking*

As was the case in the discontinuum approach, the elastic domain is restricted by both a tension cut-off in the tension area and a yield contour in the compression area, the difference being that the concepts have now been defined in the principal stress area instead of the σ, τ-area. The idea of the continuum approach has been illustrated in Figure 3.16. The effect of a crack (or shear plane) is spread out over the area that belongs to an integration point. This is referred to as the 'smeared crack concept' as opposed to the discrete crack concept. Whereas a discrete crack concept is based on crack displacements and stress displacement relations Equation (3.7), crack strains and stress-strain relations are used in a smeared crack concept. It is then obvious what the advantages of this method are: in the case of crack formation the linear stress-strain relation from Equation (3.29) should only be replaced by a stress-strain relation with regard to the cracked stage; no special interface elements or other devices are needed. An additional advantage is that there are no restrictions with regard to the direction and location of the cracks. A smeared crack can occur in the element mesh at any location and in any direction. The stress-strain relation is transformed according to standard procedures from the crack co-ordinate system to the global coordinate system and vice versa. Similar observations apply to continuum plasticity, as opposed to friction/slip models for discontinua.

For concrete, smeared crack models have been developed in many variants. For a survey refer to [27]. Since concrete is initially isotropic and masonry initially anisotropic, an extension is necessary. The variant with a so-called stress decomposition [44, 45] is particularly suitable. The overall strain is separated in an elastic part of the masonry between the cracks and a cracked part. With regard to the elastic component the orthotropic relation from Equation (3.29) can be maintained, whereas for the crack component a relation has been formed that takes into account both tension-softening and shear. After the necessary transformations, both components are combined into an overall relation for the cracked masonry. Actually this leads to a double form of orthotropy: initial orthotropic elastic behaviour is combined with orthotropy as a consequence of crack formation. The procedure has been worked out in [32, 33]. An extensive treatment is omitted here because so far all the practice-oriented case studies could be carried out by means of discontinuum models.

A second extension in comparison to concrete is that the tension cut-off and the tension-softening parameters for masonry depend on the direction. The parameters are a function of the inclination angle α between the direction of crack and joint. The tensile strength f_t and the fracture energy G_f^I are different for cracks parallel to the longitudinal joints and cracks perpendicular to the longitudinal joints, that is to say $f_t = f_t(\alpha)$ and $G_f^I = G_f^I(\alpha)$. Furthermore, the smeared crack approach implies a crack band width [27] as discretization parameter, needed to correctly carry out the translation from crack displacement to crack strain.

In comparison with the advantages mentioned with regard to the smeared crack approach, there are disadvantages among which the problems concerning reduction of the shear resistance and the fact that 'locked for' stresses are found which lead to too stiff behaviour. A survey is given in [46], in which an isotrope damage model appears to be an attractive alternative.

3.5.4 *Plasticity*

Plasticity models are suitable to describe crushing and shear in compression. The generalisation of Coulomb's friction criterion to continua leads to the Mohr-Coulomb yield criterion, the shape of which is polygonal. As an alternative non-cornered yield criterion is sometimes used, such as Drucker-Prager or Von Mises criteria. The extension to masonrylies in the addition of orthotropics. The orthotropic yield criteria according to Tsai-Hill and Hoffmann which have proven their value with regard to engineering plastics, offer good opportunities. These possibilities have not yet been worked out because the case studies considered so far, were not critical with regard to collapse due to compression.

The new trend to formulate not only the compression side but also the tension side via plasticity, occurs with both interface models (Figs 3.12a and c) and with continuity models. Feenstra [47] recently developed a Von Mises/Rankine yield criterion for concrete. An extension of this approach to orthotropics for masonry deserves attention. The approach has a lot in common with the composed yield criterion for interface elements (Fig. 3.12c). Another trend is formed by plasticity for so-called Cosserat-continua [48]. In that approach the micro-structure of the material is included into the macro-formulation. Where the prediction of shear bands is concerned, this technique shows a considerable improvement to numerical stability and objectivity. With soil the underlying granular structure is accounted for by rotations. In the case of masonry the stacking pattern (unit bond) can be applied as micro-structure.

3.6 SENSE AND NONSENSE WITH REGARD TO THE SCATTER IN MATERIAL PROPERTIES

At the start of the A33 CUR committee the new numerical methods for masonry were looked upon suspiciously by sceptics. A frequently used argument with regard to that was: 'given the large scatter of properties of masonry, an accurate numerical approach is senseless'. Apart from the fact that it is suspicious that this reasoning is obviously not applied to traditional analytical research, the following can be stated.

Indeed, there are few materials of which the variety in properties is as large as with masonry. Weather influences before and during unit laying, experience, craftsmanship and even the unit layer's mood, non-constant mortar mixes and workability and variation in unit properties are many factors that influence the quality of the finished product. In terms of percentage it are not so much the properties under compression, but especially the tensile strength, tension-softening and shear strength that are influenced. The average values of these properties are already low in comparison to, for example, concrete. Initially, it was proposed that the solution would be to neglect the tensile strength and tension-softening altogether (no-tension material). With such an approach any cavity wall would collapse and any unit wall would instantly crack. Modern slender masonry structures are placed in bending or tension, so that tension zones, bond strength and tensile-bending strength are essential in the absence of reinforcement. Only arches that involve compression only would remain intact. This partly accounts for the discrepancy between safety considerations and reality, as mentioned in Chapter 1. At present it is gradually realized that tensile strength is a property of which a lower limit should be guaranteed. Diaphragm walls [6] are a good example, where the bond tensile strength as a first priority should be guaranteed by means of quality requirements and quality control at the building site.

A second proposition subsequently was to observe the tensile strength and not the tension-softening, that is to say that after crack formation an immediate drop to zero of the stress is assumed. Such a model from a numerical point of view gives non-objective results, both in case of calculations with smeared cracks [45] and with discrete cracks [41]. The result then largely depends on the fineness of the mesh and, therefore, should be discarded. Softening is required, both on physical as well as on numerical grounds. In the case of structural concrete, it has meanwhile become accepted to rely on tensile strength and softening as point of departure in design and calculation.

In this research the scatter in material properties is approached in a more realistic way. In a number of cases, the scatter has been directly modelled by distributing the strength and softening properties at random over the element mesh, for example, by means of a standard normal distribution. This method makes sense as soon as undisturbed areas with almost uniform stress values occur, such as in long walls under restrained shrinkage. Apart from edge effects, the tension stress along the wall's longitudinal axis is nearly constant, so that in reality the weakest spots will crack. This situation can be simulated numerically with a scatter of strength values [49].

In most cases it is not necessary to apply a full scatter, since one or two imperfections are sufficient. The imperfections should be applied with some judgment. Examples are available for tension tests of reinforced [27] and unreinforced concrete [50]. The technique can be compared to experimental research in which often a notch, a reduction in cross-section or a reduced joint surface area is introduced to fix the location of the crack and the transducers. Actually, the method with interface elements is an extreme way of introducing an imperfection: crack formation (or slip) is only allowed to occur in the prescribed interface elements while cracking in the surrounding material is prohibited. There is a striking similarity with a movement joint, alternatively called expansion joint or dilatation joint.

Finally, it should be noted that the scatter and material imperfections are not important as soon as the geometry of the structure gives rise to stress concentrations. In the case of the facade with openings in Figure 3.15 the corners of the openings act as crack initiators. The stress concentration exceeds the effect of a scatter of tensile strength values.

CHAPTER 4

Evaluation and verification studies with DIANA

4.1 GENERAL

After micro-material tests have been carried out, as well as the formulation of material models and the implementation of those models within the finite element method, the next step in this research is verification and evaluation (refer to Section 1.2, framework of the research). It has to be assessed whether the numerical models and parameters can accurately enough reproduce the macro-behaviour of structures. For that reason, structures should be analysed of which either experimental or analytical solutions are known.

The verification-objects in this chapter are piers in shear and wall parts in tension. With regard to the piers, experimental data are available and for the wall parts analytical solutions function as a reference.

4.2 VERIFICATION STUDY OF PIERS IN SHEAR.

4.2.1 *Experimental results*

An extensive series of tests was carried out on piers loaded in shear by Vermeltfoort and Raijmakers [51, 52] within the framework of the B50 CUR committee. The experiments were especially designed to verify numerical models. A number of piers have been simulated numerically. These piers have a width/height ratio equal to 1 (990 × 1000 mm), composed of 18 layers of which 16 are actually loaded, and 2 layers were fixed in steel beams which make sure that the load introduction is not disturbed (Fig. 4.1). The units are of the Joosten type with dimensions of 204 × 98 × 50 mm and the mortar joints with a thickness of 12.5 mm are composed according to a cement:calcium:sand ratio by volume of 1:2:9. These materials are the same as those used in the micro-tests of Chapter 2. Apart from the piers, small companion tests were carried out to determine the micro-properties under the same circumstances.

The piers were pre-loaded by a vertical top load F_v of 30 kN, after which a horizontal load F_h was added in displacement control. After the horizontal load had been added, the top and bottom edge were kept straight, horizontal and in its place, thus resulting in a fully damped and fixed situation. Load displacement diagrams and crack patterns have been registered. Figure 4.2 shows the crack patterns of two of the

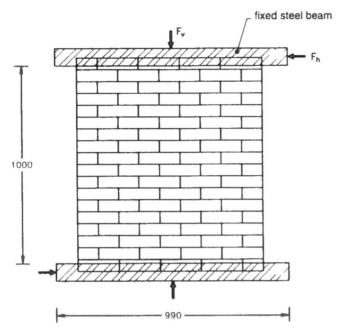

Figure 4.1. Pier with vertical pre-load F_v and increasing horizontal load F_h. The top and bottom edge were completely fixed after the application of F_v

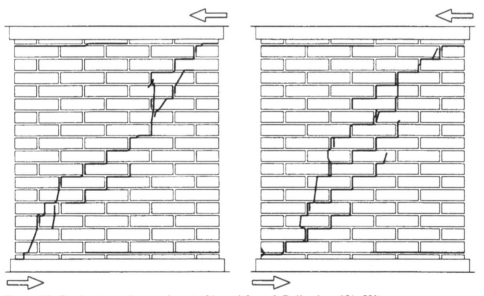

Figure 4.2. Crack patterns in experiments (Vermeltfoort & Raijmakers [51, 52]).

three tests from the series concerned (every series consisted of three identical tests so that an impression of the scatter was obtained).

The first cracks appeared to occur at a relatively low horizontal load. These are horizontal bending cracks at the bottom and top longitudinal joint, where the initial tensile stresses are at a maximum. Through redistribution the force is subsequently transferred via a compression diagonal. All of a sudden a diagonal shear crack appears when the load increases. The compression diagonal in the pier splits under lateral tension. The diagonal crack mainly runs along the vertical joints and longitudinal joints, in a stepwise manner, but crack formation also occurs to a limited extent in the units. The extent of crack formation in the units versus crack formation along the joints is dependent on the ratio of unit strength and joint strength. The diagonal crack finally leads to collapse accompanied by local crushing of the mortar or crushing of the units under compression in the bottom left and the top right corners.

4.2.2 Modelling

The units have been modelled with plane stress elements (8 elements across the length and 2 across the height of the unit) and the joints with interface elements. In the first instance, the behaviour of the units has been kept linear-elastic. As a result they cannot crack. All crack formation and slip is concentrated in the joints. As an interface model Coulomb friction was selected with the parabolic tension cut-off from Figure 3.12b (the difference in result between the parabolic and the straight tension cut-off from Figure 3.11 was only marginal for this structure) [39]. The material parameters are shown in Table 4.1. These input parameters are the result of a compromise between the results of the micro-tests from Chapter 2 and the results from the companion experiments with the piers [51, 52].

The joints have been modelled with zero thickness, according to Figure 3.5c. Initially, the accompanying stiffness values k_n and k_t were determined according to the Equations (3.14) and (3.15) and Fig. 3.7, taking into account the elastic moduli of the real joints. For those elastic moduli, the average values from the tension and compression tests were used. The values of k_n and k_t determined that way caused a too stiff behaviour of the pier. For that reason the parameters have been reduced to the values shown in Table 4.1. The linear stiffness values determined from micro-tests do obviously not correspond with the linear stiffness of the pier. This translation form micro into macro should still be clarified. Probably initial irregularities and load-spreading effects have more impact in the macro-tests than in the micro-tests. At first the vertical top load was applied and subsequently the horizontal displacement was incremented, completely in analogy with the experimental set-up.

Table 4.1. Material parameters for piers made out of Joosten clay units.

Component	Parameter	Symbol	Value	Dimension
Unit	Elastic modulus	E	16700	N/mm^2
	Poisson's ratio	ν	0.15	–
Joint	Normal stiffness	k_n	82	N/mm^2
	Shear stiffness	k_t	36	N/mm^2
	Bond tensile strength	f_t	0.25	N/mm^2
	Mode-I fracture energy	$G_f{}^I$	0.01	J/m^2
	Cohesion	c	0.375	N/mm^2
	Mode-II fracture energy	$G_f{}^{II}$	0.05	J/m^2
	Angle of internal friction	$\tan \phi$	0.75	–
	Angle of dilatancy	$\tan \psi$	0	–

4.2.3 *Numerical results*

Figure 4.3 shows the calculated relationship between the horizontal load F_h and the horizontal displacement, together with the three experimental curves. The experimental curves show a gradual hardening to a certain maximum and subsequently some softening. Near the maximum the diagonal crack suddenly appeared. One of the three experimental curves lies considerably higher than the other two, so that the validity of that result should be doubted. The numerical result shows the same trend up to the peak. After the peak the force suddenly drops off, after which another rise can be observed.

Figure 4.4 shows the deformed element meshes with increasing horizontal displacement. The incremental deformations are illustrated here, that is to say the increase of deformations in the last load increment that was carried out (the cracking activity is clearly visible then, better than in the case of the total, accumulated deformations). The deformations have been magnified, as is the case throughout this report. At first bending cracks propagate horizontal in the longitudinal joints in the top left corner and the bottom right corner (Fig. 4.4a) Subsequently a number of vertical joints and longitudinal joints start in the centre of the pier to crack (Fig. 4.4b). Next the compression diagonal suddenly splits open (Fig. 4.4c). The explosive character corresponds with the drop off in load. During this process the bending cracks temporarily become inactive: they are closing as appears from the overlapping element lines in the top left and bottom right corner. When Figure 4.4c is closely examined we can see that there occurs not just one single diagonal crack, but several diagonal cracks overlapping each other, as it were. Ultimately, the middle diagonal crack appears to attract most of the deformation (Fig. 4.4d).

In a quality sense, the similarity with the experimental results can be described as excellent. This certainly applies to the horizontal bending cracks. But also the stepwise diagonal crack with its characteristic crack widths in the vertical joints, and

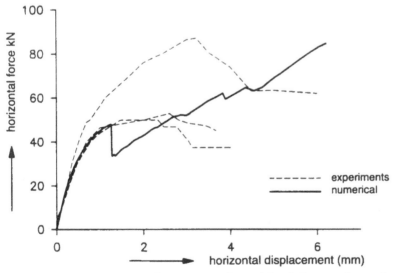

Figure 4.3. Load-displacement diagrams, experimental (in triplicate) and numerical.

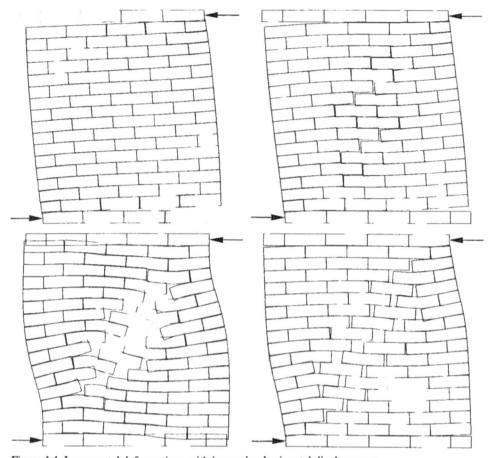

Figure 4.4. Incremental deformations with increasing horizontal displacement.

crack parallel displacements (slip) along the longitudinal joints favourably correspond with the experimental pattern (Fig. 4.2). The presence of several, overlapping stepwise diagonal cracks can also be recognised in the experiments.

In a quantitative sense, there are differences in the load displacement behaviour. Although the sudden appearance of the diagonal crack was also mentioned in the experiments, no brittle drop was observed there. The peak loads that cause the diagonal crack, do correspond well indeed. Usually, from a practical point of view, this is most important. After the drop off a rising curve is numerically observed again, while the experiments show a residual flat curve (Fig. 4.3).

4.2.4 *Evaluation of modelling aspects*

The cause of the drop off after the peak loading lies in the fact that the stresses in many vertical joints suddenly simultaneously reach the apex of the Coulomb friction criterion. Most of the vertical joints in the compression diagonal are loaded by a combination of shear stress and compressive stress. The critical stress situation, therefore, lies just left of the origin in Figure 3.12b, so that the damage begins with

Coulomb friction. During slip and cohesion-softening the stress situation appears to drift to the tension/shear area and finally the apex is reached. Gradual softening in the apex has not been taken into account: on reaching the apex the stresses suddenly reduce to zero (refer to Sections 3.3.5 and 3.3.6).

This sudden reduction of stress at integration point level accounts for the sudden load drop at structural level. Figure 4.5 shows an alternative analysis in which the stresses are not reduced to zero after the apex has been reached, but keep their level (fully plastic). The drop in load has then considerably been levelled off. The real behaviour will probably be somewhere between the two extremes. Gradual softening in the apex [37, 53] is a necessary extension of the model.

The cause of the unlimited increase of the load displacement behaviour after the peak and drop-off is mainly due to the fact that the units have been modelled purely linear-elastic. Additional calculations were made in which failure of the units in compression was modelled. For that purpose an elastic-perfectly plastic model was used with a Von Mises yield criterion and an effective compressive strength $f'_c = 12$ N/mm^2. The remaining parameters were kept the same. The load displacement diagrams are compared in Figure 4.6.

Instead of the unrestricted rising curve, the calculations with unit-plasticity show a final flat area. The units at the bottom left and the top right, on both ends of the compression diagonal, appear to have become plastic. The compressive stress in the diagonal is then limited and the load cannot increase any further. The pier appears to shear off completely now (Fig. 4.7).

Apart from a restriction of the compressive stress in the units, a restriction with regard to tension is also necessary. For a unit is always fixed between two other units, and the friction between them leads to tension stresses in the unit which in the case of linear-elastic behaviour can reach to an unjustifiable level. First an attempt was made with smeared cracking in the units [36]. These analyses proved to be unsuccessful. The combination of smeared cracking in the units with discrete crack formation in the joints led to bifurcations and convergence problems.

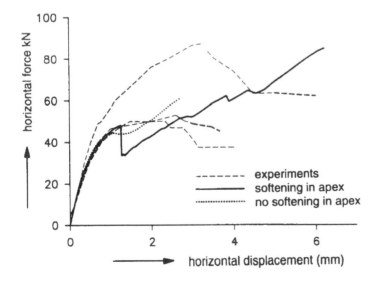

Figure 4.5. Influence of apex softening on the load displacement behaviour.

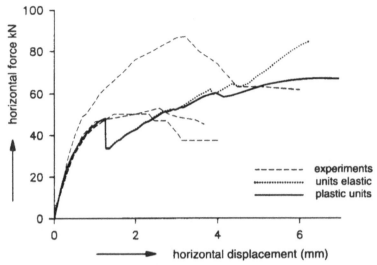

Figure 4.6. Load displacement diagrams for analysis with linear-elastic units and with a plastic restriction for failure of units in compression.

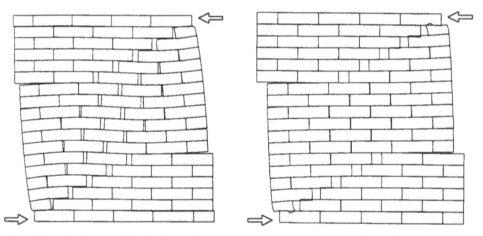

Figure 4.7. Calculation with plastic restriction for units in compression. a) Total displacements at collapse, b) Incremental displacements at collapse.

An alternative is the use of discrete cracking in the units. For that purpose vertical interface elements can be placed in the middle of the units. Then a crack does not necessarily need to run along longitudinal joint and vertical joint, but could also jump from vertical joint to vertical joint, straight through the unit. The latter was also observed experimentally. This method had been successfully applied before with piers in compression [32]. Lately, the piers in shear were also analysed by means of this method, in which at the same time apex-softening, a compression cap (Fig. 3.12c) and cap-softening were added to the interface model [53]. The quantitative results then fully coincide with the experiments, also with respect to the meas-

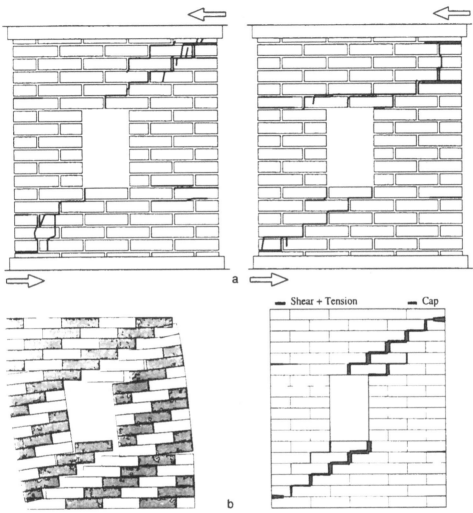

Figure 4.8. Failure behaviour of pier with opening. a) Experiments (Vermeltfoort & Raijmakers[51, 52]), b) Recent numerical result with composite yield criterion for interface elements (Lourenco & Rots [53]).

ured reaction forces in the various jacks. Figure 4.8 shows a typical result of a pier with opening.

4.2.5 *Striking influence of dilatancy*

In the previous calculations the angle of dilatancy ψ was equal to zero. This parameter is an indication of the uplift in shear (refer to Section 3.3.4, Fig. 3.10). For $\psi = 0$ the shear across a longitudinal joint is not accompanied by normal displacements. Additional calculations were carried out for tan $\psi = 0.4$ ($\psi = 22°$) among others. The other parameters were kept equal and the units were assumed to be linear-elastic. The load-displacement diagrams are compared in Figure 4.9.

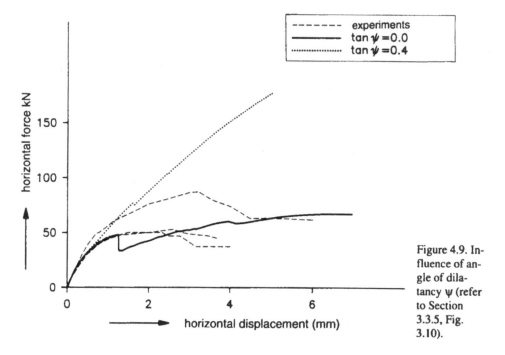

Figure 4.9. Influence of angle of dilatancy ψ (refer to Section 3.3.5, Fig. 3.10).

The influence of ψ is extraordinary large. Whereas a peak load and considerable divergence of the curve was found for tan ψ = 0, no peak load whatsoever was predicted for tan ψ = 0.4. There is an unrestricted increase in the load and the behaviour is extremely stiff.

This can be explained by the fact that the pier was confined by the steel beams. The steel beams at the top and bottom were straight, horizontal and fixed in their place. Dilatancy in a longitudinal joint is therefore obstructed by the boundary conditions. The material cannot 'get away'. The dilatancy in the longitudinal joints should therefore be compensated by compression in the units. The horizontal pulling apart of the units in case of expanding longitudinal joints is accompanied by a wedging effect. The consequences are extremely high vertical compressive stresses and a very stiff behaviour. In the extreme calculations with ψ = 0 slip can occur unrestrictedly along the longitudinal joint. This situation is similar to that of perfectly smooth units which have been laid perfectly horizontal. Figure 4.10 shows the deformations with and without dilatancy.

In reality there always exists a certain roughness and non-smoothness of both unit and bond surface. The micro shear tests from Section 2.5.8 give values for tan ψ between 0.6 and 0, depending on the type of unit and mortar. Calcium silicate units usually have a smoother surface than clay units and indeed show lower values. Even more important, however, is the decrease of the angle of dilatancy with increasing compressive stress (Fig. 2.30) and increasing slip. This should be translated into an extension with softening on the dilatancy. Although such an extension of the numerical model is possible, it has not yet been carried out. So far the input for ψ was often either very low or tan ψ = 0, so that the results form a safe lower limit. The discussion will be continued in Section 4.3.6.

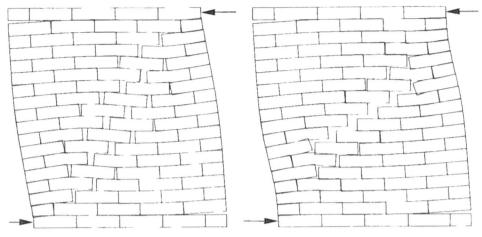

Figure 4.10. Deformed element meshes. a) tan ψ = 0.0. Units slide across each other without uplift in the longitudinal joint, b) tan ψ = 0.4. Uplift in cracked longitudinal joints is visible.

4.2.6 *Conclusions*

The numerical model existing of a combination of Coulomb friction with cohesion-softening and discrete cracking with tension-softening, has been verified by means of experiments on piers in shear. In a qualitative sense, the results show very good agreement with the experiments in shear. The pattern of horizontal bending cracks and diagonal zigzag cracks can accurately be simulated. In a quantitative sense, it is possible to predict the behaviour up to the maximum load accurately enough.

Initially, differences were found in the post-peak behaviour. Improvements were obtained by non-linear modelling of not only the joints, but of the units as well. The evaluation of a number of numerical aspects, among which apex-softening, addition of a compression cap and addition of discrete cracks in the units recently resulted in a model in which the results, also from a quantitative point of view, accurately correspond [53].It is surprising that the dilatancy in longitudinal joints plays a decisive role. This parameter which, like softening, hardly occurs in masonry literature, appears to dominate the strength and stiffness of 'locked up' structural parts. The outcome of the verification study in this section justifies application of the models in practice-oriented case studies.

4.3 EVALUATION STUDY OF WALL PARTS IN TENSION

4.3.1 *From micro to macro via numerical simulations*

In many practical studies it is not so much the micro-properties of the components that are needed, but the macro-properties of a wall part. These macro-properties of the composite can be numerically deduced from the underlying properties of the components (unit, joint and area of adhesion) and the stacking pattern (masonry bond). In this section an evaluation takes place of the relation between micro and macro for wall parts in tension. After a comparison with analytical results, a demonstration will be given of the fact that numerical methods can predict the effect of the (many) influencing parameters on the composite behaviour.

The macro-properties can subsequently be used for global analyses purposes, for example in the case study of cracking behaviour of walls under restrained shrinkage (Chapter 6), where the behaviour of masonry in horizontal tension is of importance. This section's procedure can be considered as the step 'from unit to wall part', whereas the case study in Chapter 6 can be marked with the motto 'from wall part to wall'.

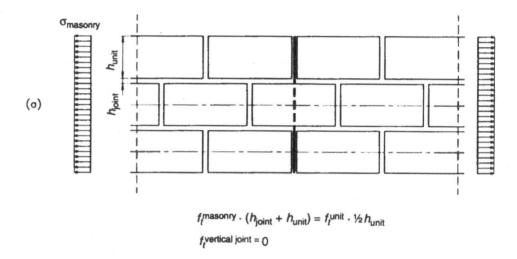

$$f_t^{masonry} \cdot (h_{joint} + h_{unit}) = f_t^{unit} \cdot \tfrac{1}{2} h_{unit}$$

$$f_t^{vertical\ joint} = 0$$

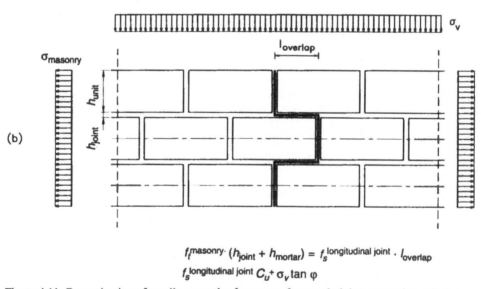

$$f_t^{masonry} \cdot (h_{joint} + h_{mortar}) = f_s^{longitudinal\ joint} \cdot l_{overlap}$$

$$f_s^{longitudinal\ joint}\ C_u + \sigma_v \tan \varphi$$

Figure 4.11. Determination of tensile strength of masonry from underlying properties and dimensions of the components (unit and joint) and the stacking pattern (unit bond). Taken from Schubert et al. [54, 55]. Assumption: pre-cracked vertical joints. a) Tension crack in unit is normative, b) Shear slip along longitudinal joint is normative.

4.3.2 *Analytical reference data*

Analytical observations of the behaviour of masonry in tension have been recorded by Schubert and his staff members [54, 55] (Fig. 4.11). They assume that vertical joints are weak, pre-cracked spots which cannot take tension. Then there are two possible failure mechanisms: a) Failure of the unit due to tension, b) Shear slip along the longitudinal joint. The masonry tensile strength has been determined as the lowest value of these two possible forms of failure:

– Tension crack in the brick:

$$f_{t;\text{masonry}} = f_{t;\text{unit}} \frac{h_{\text{unit}}}{2(h_{\text{unit}} + h_{\text{joint}})} \tag{4.1}$$

– Shear slip along longitudinal joint:

$$f_{t;\text{masonry}} = (c_u + \sigma_v \tan \phi) \frac{l_{\text{overlap}}}{h_{\text{unit}} + h_{\text{joint}}} \tag{4.2}$$

where $f_{t;\text{masonry}}$ is the tensile strength of the masonry, h_{unit} is the thickness of the unit, h_{joint} is the thickness of the joint, l_{overlap} is the overlapping length of the units (dependent of the bond) and σ_v is the vertical compressive stress from the top load. These formulae can simply be deduced by equilibrium, assuming that there is a uniform distribution of tensile stress and shear stress. In reality there are stress peaks present and due to gradually progress of softening, a different picture will be obtained. This additional information can be gained by a numerical study which will also give information about the stiffness and deformation capacity of the wall part.

4.3.3 *Numerical modelling*

A representative piece of masonry from a large wall is considered which is subjected to a horizontal tensile load, as illustrated in Figure 4.12. The example concerns glued calcium silicate elements with a length of 900 mm, a height of 600 mm and a width of 100 mm. These are gross dimensions including the thickness of the thin glue joints of 3 mm for a vertical joint and 2 mm for a longitudinal joint. The elements are placed in half-unit bond. Because of the symmetry only two half unit-thicknesses plus joint need to be observed in vertical direction. Both the top and the bottom edge of the modelled part are kept straight and horizontal (u_y = constant). The bottom edge is fixed in vertical direction, while the top edge can move in vertical direction. In longitudinal direction two and a half stretch has been modelled. In the middle a potential crack is placed along the vertical joints and longitudinal joint (toothed shape). This modelling implies an assumption of a crack distance of 2250 mm. The left and right edge of the modelled part remain straight (u_x = constant). With a length of two and a half stretch, the distance between the edges and the crack is considered large enough to avoid a possible edge disturbance on the cracking process.

Interface elements have been placed in all the vertical joints and longitudinal joints. The interface elements at the location of the potential crack are assumed to be non-linear with Coulomb friction and tension-softening according to Section 3.3.6 (Fig. 3.11). The other interface elements have been kept linear-elastic. The parame-

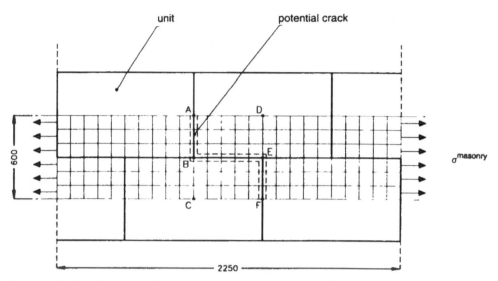

Figure 4.12. Modelled piece of masonry taken from an infinite wall under horizontal tension. Dimensions correspond with calcium silicate elements of 900 × 600 mm. In the middle a potential crack along longitudinal joints and vertical joints is adopted. Owing to symmetry the four edges of the element mesh remain straight.

Table 4.2. Material parameters for detailed study of wall parts that are placed in tension. Values relate to glued calcium silicate elements.

Component	Parameter	Symbol	Value	Dimension
Units	Elastic modulus	E	5000	N/mm²
	Poisson's ratio	v	0.2	–
	Tensile strength	f_t	1.3	N/mm²
Vertical joint	Normal stiffness	k_{nn}	333	N/mm²
	Shear stiffness	k_{tt}	139	N/mm²
Longitudinal joint	Normal stiffness	k_{nn}	500	N/mm²
	Shear stiffness	k_{tt}	209	N/mm²
Joint	Tensile strength	f_t	0.5	N/mm²
	Cohesion	c_u	0.75	N/mm²
	Angle of internal friction	tan φ	0.75	–
	Angle of dilatancy	tan ψ	0.2	–
	Mode-I fracture energy	G_f^{I}	0.01	J/m²
	Mode-II fracture energy	G_f^{II}	0.05	J/m²

ters for glued calcium silicate unit were not fully known when these calculations took place. Based on available data [55-57], assumptions have been made. These are indicated in Table 4.2. The initial stiffness values have been determined from the joint's elastic properties including areas of adhesion, according to $k_n = E/h_{\mathrm{joint}}$ and $k_t = (E/(2(1 + v)))/h_{\mathrm{joint}}$ in which it is assumed that $E = 1000$ N/mm², $v = 0.2$ and h is the thickness of the glued vertical or longitudinal joint.

4.3.4 *Results*

Figure 4.13 shows the calculated relationship between the external tensile stress on the masonry $\sigma_{masonry}$ and the shrinkage strain. (Actually, it was not 'shrinkage' that was imposed, but elongation. This elongation, however, has been re-adjusted to a shrinkage strain by dividing it by the length (2250) of the wall part considered. To engineers shrinkage strain is a recognisable conception design and analysis). The diagram shows three peaks. The first peak corresponds with the cracking of the vertical joints along A-B and E-F in Figure 4.12. These are pure tension cracks (Mode-I). The second peak corresponds with the formation of a shear crack along the longitudinal joint B-E (Mode-II). The third peak corresponds with a tension crack in the unit, in the extended part of the cracked vertical joint, path B-C or D-E in Figure 4.12. This crack has not been modelled as such (no potential crack has been placed in the unit), but the external stress that it is caused by, can be determined indicatively according to Equation (4.1) as half of the unit tensile strength which is 0.83 to 1.45 N/mm^2 for calcium silicate unit elements [56]. Here a value of 1.3 N/mm^2 was used. Brittle snap-back behaviour occurs after all three peaks. The behaviour after the second peak has completely been traced by means of a special arc length procedure (refer to Section 3.2.2). The snap-backs after the first and third peak have not been included in Figure 4.13. Instead, it is shown diagrammatically that the stress suddenly drops off as the deformation increases. Snap-back behaviour seems strange at first, but is also found experimentally when special feedback equipment is applied (refer, for instance, to [13]). In fact this implies that the piece of masonry relaxes on both ends while the crack propagates. When an inadequate control parameter is used, then an explosive crack propagation occurs in reality and in numerical analysis. This phenomenon has to do with the ratio between the amount of elastic energy that is stored in the structure, and the fracture energy dissipated during crack propagation. Snap-back behaviour is more pronounced with masonry than with concrete because the fracture energy in masonry is a factor of 5 to 10 lower than for concrete. Besides, with tension cracks it occurs to a **greater** extent than with shear cracks because G_f^I is only approximately one fifth of G_f^{II}.

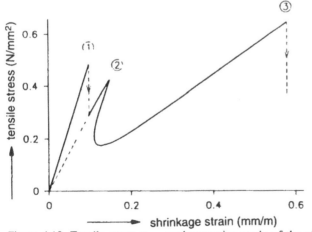

Figure 4.13. Tensile stress σ_{unit} against tension strain of the piece of masonry from Figure 4.12. Peak 1: vertical joints are cracking. Peak 2: longitudinal joint is cracking. Peak 3: unit is cracking.

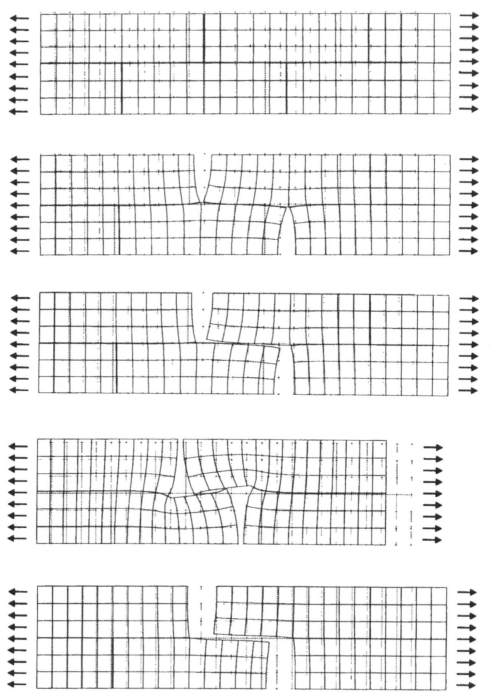

Figure 4.14. Deformation process of wall part under tension. a) Initial, vertical and longitudinal joints still uncracked (before Peak 1 in Fig. 4.13), b) Cracked vertical joints (immediately after Peak 1), c) Longitudinal joint partially cracked (just before Peak 2), d) Longitudinal joint cracks completely (incremental deformations, snap-back after Peak 2), e) Residual stage with dilatancy in longitudinal joint (just before Peak 3).

The development of the cracking process has been visualised in Figures 4.14 and 4.15, which respectively represent the deformations and principal stresses at increasing stages of the loading process. It is obvious that the crack in the vertical joints is a pure tension crack (Mode-I) (Fig. 4.14b). After the cracking of the vertical joints the principal tensile stress winds its way through the piece of masonry (Fig. 4.15b). The principal tensile stresses make only a small angle with the longitudinal joint. The longitudinal joint is therefore submitted to a high shear stress and the resulting crack along the longitudinal joint can be recognised as a shear crack (Mode-II) (Fig. 4.14c and 4.14e). All integration points along the vertical joint have reached the tension cut-off, while Coulomb friction occurs in all integration points along the longitudinal joint.

In Figure 4.14d the propagation of the active shear crack can be recognised, while the end faces relax and relieve the vertical joints temporarily. The figures also illustrate the dilatancy in the longitudinal joint. Since slip is accompanied by a normal displacement (tan $\psi = 0.2$), and since the lower and upper edge are forced to remain straight, the unit wedges itself until it is locked in (Fig. 4.14e). As the slip increases, the vertical compressive stress on the longitudinal joint increases as well (Fig. 4.15c). As a result, the external tension stress can increase until the unit will crack (Peak 3).

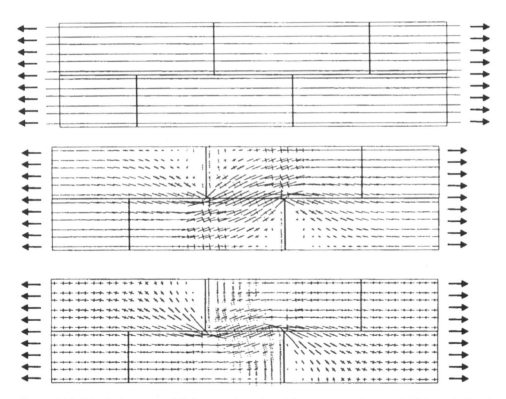

Figure 4.15. Principal stresses (full line: tension, dotted line: compression). a) Initial, vertical and longitudinal joints still uncracked, b) After the cracking of the vertical joints, c) Final stage. Compressive stresses across the longitudinal joint illustrate the wedging effect.

4.3.5 *Comparison with analytical formulations*

A comparison of the numerical analysis with Schubert's analytical approach (refer to Section 4.3.2, Eqs 4.1 and 4.2) give the following results for masonry tensile strength of glued calcium silicate elements:

analytical: $f_{t;masonry}$ = minimum (0.65, 0.56) = 0.56 N/mm^2

numerical: $f_{t;masonry}$ = 0.42 N/mm^2

The numerical value corresponds with Peak 2 in Figure 4.13. The analytical approach, which is widely used, appears to be 25% more optimistic than the numerical analysis. This is caused by the fact that a uniform distribution of the shear stress across the longitudinal joint is assumed, whereas the numerical method takes into account non-uniform distributions and softening. Due to the stress peaks on the left and right ends of the overlap, the shear crack already occurs at a lower external tension stress. The numerical result seems plausible. It is not yet possible to carry out a comparison with experiments since there are no data available in literature.

A second aspect of the numerical analysis concerns the fact that insight is gained in the deformation behaviour, whereas the analytical approach exclusively refers to strength. In the analytical approach a zero stress is assumed in vertical joints, in other words, the first peak is neglected. From the present analyses we can see that this gives a considerable reduction in stiffness; the slope of the second ascending part of the curve in Figure 4.13 amounts to only 63% of the slope of the first ascending part of the curve. This reduction is extremely relevant to the cracking behaviour under restrained shrinkage. A lower stiffness of the wall has indeed a positive effect because the wall is more able to withstand prescribed deformations. This only applies, of course, when the local crack width in the vertical joints remains acceptable.

4.3.6 *Influence of dilatancy*

A third interesting aspect is the role of the angle of dilatancy ψ, which is lacking in analytical approaches. The calculations have been repeated for tan ψ = 0.75 and tan ψ = 0, instead of tan ψ = 0.2. Figure 4.16 shows that this has a considerable influence on the behaviour, especially after the peak. (In this figure the first phase in which the vertical joints crack, is represented diagrammatically with a dotted line). When the dilatancy is zero no force is rebuilt after shear slip, whereas an associated flow rule with tan = tan ψ = 0.75 results in a very strong and stiff behaviour. The deformations for ψ = 0 (Fig. 4.17c) illustrate that slip can occur unrestrainedly along the longitudinal joint. When tan ψ = 0.2 (Fig 4.17b) and, to a larger extent when tan ψ = 0.75 (Fig. 4.17a), we can observe the stiffening 'wedging effect' under the given boundary conditions (top and bottom edge remain straight because of the symmetry in an infinitely large wall).

The extent of the wedging effect will depend on the roughness along unit and longitudinal joint.

On average, the micro-shear experiments result in tan ψ = 0.2 for the relatively smooth calcium silicate unit, and tan ψ = 0.35 for the relatively rough clay unit, but these values tend to reduce when the confining normal stress increases. Furthermore,

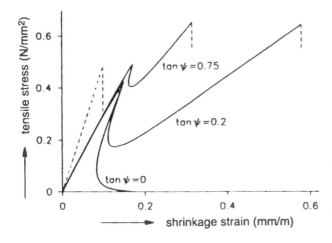

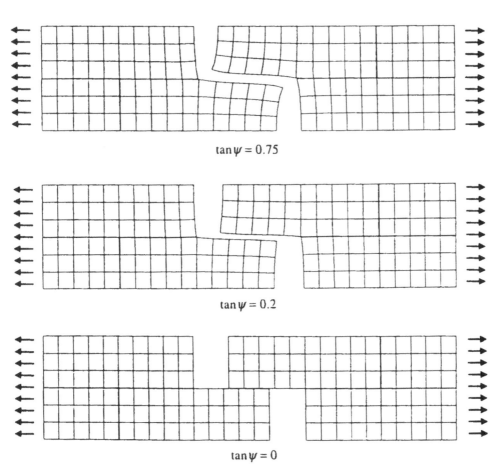

Figure 4.16. Influence of angle of dilatancy ψ (refer to Fig. 3.10) on the tensile stress-strain diagram of element-sized calcium silicate masonry. The higher ψ, the more wedging effect and the stiffer the residual behaviour.

tan ψ = 0.75

tan ψ = 0.2

tan ψ = 0

Figure 4.17. Influence of angle of dilatancy ψ (refer to Fig. 3.10) on deformations in final stage. When ψ = 0 there is no resistance against slip along the longitudinal joint. The higher ψ, the more wedging effect.

it is plausible that the dilatancy decreases with increasing shear displacement. For material with a loose grain will be 'ground', so that the surface becomes smoother as the shear increases. This softening on the dilatancy will be the subject of subsequent research. The crucial role of dilatancy for these wall parts does not only apply at detailed level, but has also been proven at structural level piers in shear, refer to Section 4.2.5), and is therefore one of the eye-openers of this structural masonry research.

4.3.7 *Influence of the type of masonry (calcium silicate unit/clay unit)*

After evaluation and comparison with the analytical results, numerical extrapolations have been made to other types of masonry. The purpose is to determine the influence of properties and dimensions of the components on the behaviour of the composite. The preceding analysis with element size calcium silicate unit functions as a reference.

Figure 4.18 shows the predicted behaviour of Vijf Eiken clay units [58]. For the glued elements, the small thickness of the joint could be neglected. For these clay units that are laid according to the standard procedure with mortar, this is not the case. The joint thickness is 10 mm and the clay units has dimensions of $210 \times 51 \times 100$ mm. The modelling of unit-area of adhesion-mortar from Figure 3.5a was used. The parameters are shown in Table 4.3.

The most clear difference is that the second part of the curve in the diagram runs on even further in comparison to the second part of the curve with element size cal-

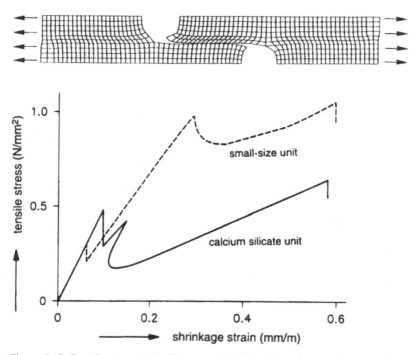

Figure 4.18. Tensile stress-strain diagram and deformations for a wall part made out of small-size Vijf Eiken clay units. Comparison with a wall part made out of element size calcium silicate units.

Table 4.3. Material parameters for masonry made out of small-sized Vijf Eiken clay units.

Component	Parameter	Symbol	Value	Dimension
Units	Elastic modulus	E	6000	N/mm^2
	Poisson's ratio	ν	0.1	–
	Tensile strength	f_t	2.5	N/mm^2
Mortar	Elastic modulus	E	1000	N/mm^2
	Poisson's ratio	ν	0.2	–
Adhesion area	Normal stiffness	k_n	10^6	N/mm^2
	Shear stiffness	k_t	10^6	N/mm^2
	Bond tensile strength	f_{tb}	0.3	N/mm^2
	Cohesion	c_u	0.6	N/mm^2
	Angle of internal friction	$\tan \phi$	0.75	–
	Angle of dilatancy	$\tan \psi$	0.1	–
	Mode-I fracture energy	G_f^{I}	0.01	J/m^2
	Mode-II fracture energy	G_f^{II}	0.05	J/m^2

cium silicate units. Apart from differences in material properties, this can be explained by the greater length/height ratio of small-size units. The overlapping length of the units has increased more than the height, so that a higher resulting shear strength across the longitudinal joint can be mobilised. This effect is also present in the analytical formulae (Eq. 4.2). small-size masonry has a larger reserve after the vertical joints have cracked, than element-size masonry.

4.3.8 *Influence of dimensions of units, blocks and elements*

In the following variation, the type of material remained constant (calcium silicate unit), but the dimensions of the components were varied [58]. In practice the following components are used: glued elements ($900 \times 600 \times 100$ mm), glued blocks ($440 \times 200 \times 100$ mm) or small-size units ($210 \times 51 \times 100$ mm) with mortar. The same parameters have been used for the blocks, elements and units. For the glued joints between the blocks, the same parameters were used as for the glued joints between the elements. Mortar and area of adhesion for the small-size masonry are characterised with E_{joint} = 2500 N/mm^2, ν_{joint} = 0.2, f_t = 0.1 N/mm^2, c_u = 0.2 N/mm^2, G_f^{I} = 0.0033 N/mm, G_f^{II} = 0.017 N/mm. The result has been illustrated in Figure 4.19. For both the blocks and the small-size units (also the small-size clay unit from the previous section), as well as for the elements, a wall part of two and a half stretch with a potential crack in the middle was adopted (Fig. 4.12). The lengths of the three modelled wall parts are therefore different. Since the shrinkage strain has been represented on the horizontal axis in the diagrams instead of the elongation, an objective comparison is possible after all. To interpret the post cracking behaviour, it should be noted that the assumed crack distance is different in all three cases.

From Figure 4.19 it can be concluded that the components with block size turn out stronger and that components with small-size are less strong than the components with element size. With regard to 'strength' the level of Peak 2 was used, corresponding with shear failure across the longitudinal joint. The differences can again

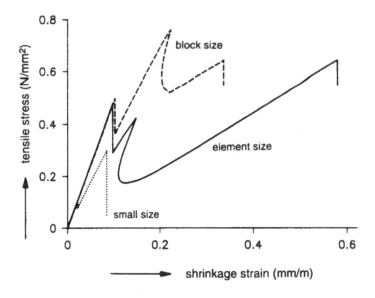

Figure 4.19. Influence of component dimensions on tensile stress-strain diagram for calcium silicate masonry. Glued element size, glued block size, small-size with mortar.

be partially explained by a change in the ratio between the overlapping length and the height of the component. Furthermore, in the case of small-size calcium silicate masonry, the relatively low adhesion properties are responsible for the decrease in strength.

The initial stiffness values of the three composites (slope of first ascending part of the curve) are similar. This slope nearly equals the value of the elastic modulus of 5000 N/mm². The effect of the thin joints on the elastic modulus of the wall part is negligible, which corresponds with results from simple hand calculations. After the vertical joints have cracked, the stiffness of the element size components appears to be a slightly lower than that of the block- and small-size components (slope second ascending part of the curve in Fig. 4.19). Here again, the ratio between the overlapping length and the height of the component is decisive.

4.3.9 *Influence of top load*

Due to dilatancy, a confining pressure develops in a confined situation. Often a confining compressive stress is already present due to the top load. This vertical pre-stress allows a higher shear stress across the longitudinal joint because of Coulomb friction, and that way the tensile strength of the masonry increases according to Equation (4.2). This has been verified numerically. An additional calculation for the calcium silicate elements has been carried out with a top load of 0.5 instead of 0 N/mm².

Figure 4.20 shows the calculated increase in strength, which is 57% ($\sigma_{masonry} = 0.66$ versus 0.42 N/mm²). Equation (4.2) results in an increase of 50% ($\sigma_{masonry} = 0.84$ with pre-stress 0.5 versus 0.56 N/mm² with pre-stress 0). There is a great similarity. The marginal difference can be explained by the fact that cohesion-softening and dilatancy are taken into account in the numerical calculation whereas this is not the case in the analytical formula.

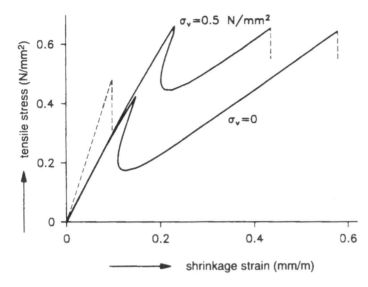

Figure 4.20. Influence of pre-stress σ_v on the tensile stress-strain diagram for glued element-size calcium silicate masonry.

In the presence of a top load it should be noted, however, that the tensile strength of the unit can become decisive according to Equation (4.1). The unit can collapse due to tension before the longitudinal joint collapses due to shear. The reserve in the masonry tensile strength with the presence of a top load is limited by Equation (4.1).

4.3.10 *Influence of open head joints*

Crack formation in vertical head joints reduces the stiffness considerably. In this example the slope of the second part of the curve in Figure 4.13 is only 60% of the slope of the first part of the curve. Because of this reduction in stiffness, the research centre of the industry (*Research Centrum Calcium Silicate Industry*) has posed the idea thought that masonry with open vertical joints might be interesting with a view to deformational behaviour under restrained shrinkage (Berkers & Rademaker [56]). The more flexible the wall, the more deformation it is able to withstand without crack formation. As stated before, the accompanying gaps (crack widths) that occur in the vertical joints should remain within certain margins, especially when there are layers of plaster present in which cracks become quickly visible. To quantify this, the calculations have been repeated for masonry that has open vertical joints, that is to say that all the vertical joints in Figure 4.12 from the start of the calculation have a tensile strength and stiffness equal to zero. The parameters and the location of the potential crack face along the longitudinal joint have not changed in relation to the original calculation. Figure 4.21 shows the external tensile stress-strain diagram and Figure 4.22 gives an impression of stresses and deformations.

The failure behaviour stays more or less the same because the shear mechanism along the longitudinal joint does not change. The initial stiffness is now only 35% of the original stiffness whereby all the vertical joints were filled (first part of curve in Fig. 4.13) and only 56% of the stiffness where only one vertical joint is open (second part of curve in Fig. 4.13, valid for assumed crack distance of 2250 mm). Whether

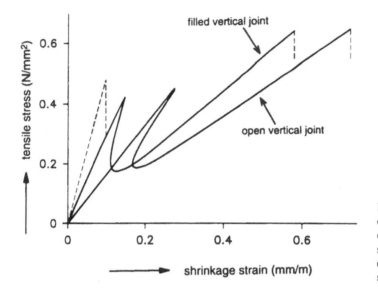

Figure 4.21. Influence of open vertical joints on the tensile stress-strain diagram of glued element size calcium silicate masonry.

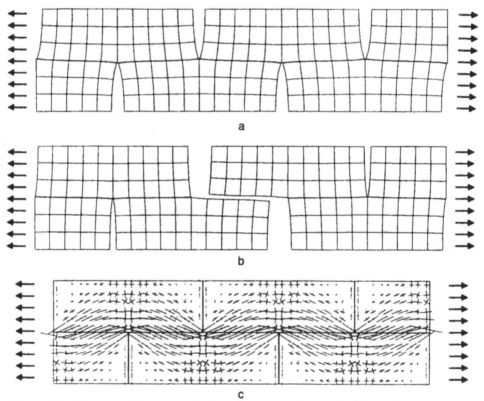

Figure 4.22. Masonry with open vertical joints; element size glued calcium silicate masonry. a) Deformations in initial, linear-elastic stage, b) Deformations in cracked final stage, c) Principal stresses in initial, linear-elastic stage.

joints are filled or not has a very strong influence on the elastic modulus of the wall. In this example the elastic modulus of the wall with filled-in vertical joints is 4910 N/mm^2 (nearly equal to the introduced elastic modulus of 5000 N/mm^2 of the calcium silicate elements), while the elastic modulus of the wall without vertical joints is 1650 N/mm^2.

A certain amount of shrinkage should be taken into account in the designs made by the construction industry. Usually 0.2 mm/m is assumed. In case of such a shrinkage, the maximum gap between glued calcium silicate unit elements with open vertical joints according to this calculation would be 0.177 mm [41]. When this value is judged in view of visual hindrance and the necessary elasticity of a possible layer of plaster, it should be realised that it is not one full-depth crack, but only a local gap in a vertical joint. The value applies to the middle of the vertical joint, while the gap at the bottom and top side of the vertical joint is zero.

4.3.11 *Conclusions*

By means of numerical micro-studies it was possible to gain insight into the behaviour of wall parts under tension. It has been proven that both softening under tension and shear as well as dilatancy play a major role in the prediction of crack formation in vertical joints, longitudinal joints and units.

With respect to analytical methods, the simulations give additional information with regard to strength, stiffness and deformational behaviour and local gaps in vertical joints. It has been proven that the macro-properties of the composite can be deduced from the underlying properties of units and joints and the stacking pattern. In this way a contribution is made in finding an optimal preparation for the composite, dependent on the functional requirements. This has been illustrated by means of a number of examples. One interesting example is the masonry with open vertical joints made out of calcium silicate unit elements, which has a positive influence when the functional requirement is aimed at following the imposed deformations (low stiffness).

CHAPTER 5

Numerical studies with UDEC

5.1 DISTINCT ELEMENT METHOD

The UDEC computer program (Universal Distinct Element Code) has especially been developed for numerical research of dynamic mechanics problems such as earth and rock slidings, vibrations and flow. The program can also be used for (quasi) static mechanics problems. The kinetic energy in the mechanics model is then eliminated by means of a damping device.

The two-dimensional program is based upon the distinct element method (DEM). An initial version of the distinct element method was published in 1971 by Cundall [59]. In 1980 the UDEC program was released which was based on this method. Subsequently the program has been further developed and new versions appeared. The calculations within the framework of the A33 CUR 'Numerical Masonry Mechanics' committee were carried out by means of the UDEC 1.63 version. For further details refer to the progress reports [60, 61].

This program's framework differs from the finite element method (FEM). The basic idea behind the distinct element method is that local fracture only occurs in the interfaces, and non-linear behaviour in the blocks. The main features of UDEC are:
– Division of the structural model into blocks and interfaces;
– Geometrical non-linearity;
– Explicit dynamic solution procedure.

5.1.1 Division into blocks and interfaces

By dividing the structural model, the input of which is a large block, into pieces, a number of smaller blocks arise. The blocks may be arbitrary geometrical polygons, and may differ among each other in shape and size. The transformation from one block into the other is called the interface which has zero thickness. The blocks can be divided into three types, namely:

1. *Undeformable blocks.* The geometry of these blocks does not change as a result of the load. The blocks are able to withstand two independent translations and one rotation. For structures where relatively low stresses and deformations occur in the blocks, and the non-linear behaviour especially occurs in the interfaces, a modelling in undeformable blocks and interfaces is realistic.

2. *Deformable blocks.* In the deformation of the blocks a distinction is made between single and multiple deformability. With a single-deformable block, two inde-

pendent constant normal strains and a constant shear strain are possible besides the two independent translations and the rotation of the entire block that occur due to the deformations of the interfaces.

A multiple-deformable block is divided into a number of triangular zones. These zones are then the continuum elements as they occur in the finite element method. For each separate zone the strain is determined. Per zone the strain, and as a consequence also the stresses, are constant. The stress-strain diagram can be assumed linear elastic or elasto-plastic.

3. *Interfaces.* In the interfaces the blocks can be connected to each other by means of contact points. Contact points are created at corners of blocks and at corners of zones which are located on the outside perimeter of the blocks (Fig. 5.1).

Per contact point two spring connections have been applied which can transfer either a normal force or a shear force from the one block onto the other block (Fig. 5.2). The force-deformation diagram that results from the springs is checked on the bases of the constitutive model. Instead of a force-deformation diagram for a spring, a stress-deformation diagram can also be made by taking into account a proportional interface length for each contact point. This proportional interface length is called contact length.

Since the interface has zero thickness, the one block will penetrate the other block under a normal compressive force. This penetration only has a mathematical significance. The relative penetration is restricted and can be influenced by the stiffness of the springs. A larger stiffness results in a smaller relative penetration but does lead to an increase in the calculation time.

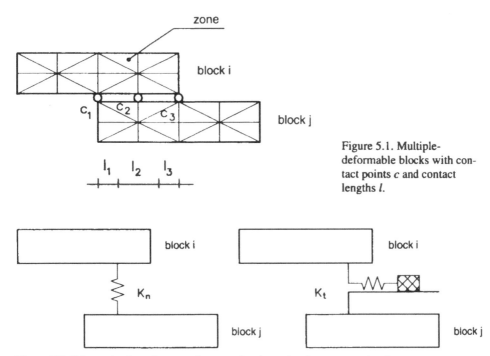

Figure 5.1. Multiple-deformable blocks with contact points c and contact lengths l.

Figure 5.2. Schematization of a normal connection k_n and a shear connection k_t.

5.1.2 *Geometrical non-linearity*

A characteristic property of UDEC is that the influence of the blocks' displacements is immediately included in the calculations. The occurring displacements can be a consequence of shear and the opening up of the interface. Due to shear and the opening up of the interface, it is possible that blocks which originally were adjacent to each other, can partially or entirely become loose from each other. It is also possible that new contact points are formed between blocks that originally were not adjacent to each other.

Because of physical reasons, the corners of the blocks were rounded off. This prevents corners from catching on to each other as a result of which an unrealistic strength could arise. The length of the rounding off usually amounts to a low percentage of the average length of the sides of a block. The rounding off of the corners is considered to be permissible since corners of blocks in practice frequently crush under an increasing stress, and surely in case of one block rotating with respect to the other.

5.1.3 *Explicit dynamic solution procedure*

In the UDEC program the explicit solution procedure is used. In this procedure the motion equations (displacement, speed and acceleration) are set up for each time increment. This takes place for each block.

In Figure 5.3 the explicit solution procedure for undeformable blocks is shown diagrammatically. At time $t = 0$ Block 1 is loaded with mass m by means of a force F. This leads to an acceleration $a = F/m$. The incremental vertical displacement after time step Δt amounts to Δw. Block 1 is connected to Block 2 by means of springs with a stiffness K. The incremental displacement Δw from Block 1 results in forces inside the springs of a magnitude of $F = K\Delta w$, which besides the reaction force F, act as reaction forces on Block 1 in the next time increment and as action forces on Block 2. Subsequently, the acceleration of Block 1 and 2 is determined.

The resulting displacements are then translated into spring forces again. This way all the blocks in the model will get a turn and also the boundary conditions of the model can be taken into account. The same principle also applies to deformable blocks, whereby the deformation of the blocks is included in each time increment.

Numerical stability and convergence to the correct final result are only possible if

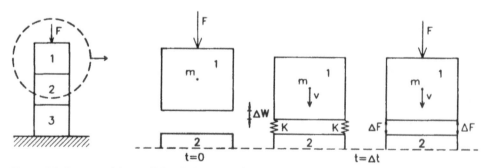

Figure 5.3. Scheme of the explicit solution procedure.

the selected time increment is small enough. In the UDEC program the time increment is automatically determined on the basis of the speed by which information runs through the model. Then the time increment is chosen so small that for each time increment the information of a block does not go further than the immediately adjacent blocks. Testing of the constitutive criteria is carried out on each block in each time increment.

5.1.4 *Constitutive models*

The original UDEC program was based on the plane straits situation. In the UDEC 1.63 version it is possible to give the stress perpendicular to the plane of the structure a constant value. The plane stress situation is then obtained by giving the stress perpendicular to the plane of the structure a value of zero.

The available non-linear models include brittle collapse under tension and shear. After the admissible tension and shear stress have been reached, the stress immediately drops off to zero. This implies that there is neither tension-softening nor cohesion-softening present.

In principle, two constitutive models are suitable for the blocks, namely a linear-elastic model and a non-linear model according to Mohr-Coulomb's failure criterion with a tension cut-off. The material parameters for the linear-elastic model are the compression modulus, the shear modulus and the weight by volume. The compression and shear modulus are a function of the elastic modulus and Poisson's ratio and have been formulated as follows:

Compression modulus: $K = E/(3(1 - 2v))$
Shear modulus: $G = E/(2(1 + v))$

Besides the parameters of the linear-elastic model the following parameters apply for the non-linear model: the cohesion, the angle of internal friction, the angle of dilatancy and the uni-axial tensile strength.

The blocks in a masonry model are connected by means of an interface. This interface can have two constitutive models, namely the continuously yielding model and an elasto-plastic model according to Mohr-Coulomb's failure criterion with a tension cut-off. In the continuously yielding model no tensile stresses are permitted in the interface, as a result of which the model is less suitable for simulation of masonry structures. The parameters for the elasto-plastic model are: the normal stiffness, the shear stiffness, the cohesion, the friction coefficient, the angle of dilatancy and the tensile strength. If in a numerical calculation the tensile strength or shear strength are exceeded at a certain location, then the tensile strength and cohesion are reduced to zero at that location. Shear stresses can only be transferred when there is a compressive stress perpendicular to the plane of the interface. The maximum transferable shear stress amounts to: $\tau = c - \sigma_n \tan \phi$.

5.2 EVALUATION STUDY OF NARROW PIER

By means of the UDEC computer program numerical calculations have been made especially for piers. This section discusses an evaluation study of a narrow pier. The next section will deal with a wide pier with opening, of which also experimental verification data are available.

5.2.1 *Modelling*

Figure 5.4a shows the geometry of the narrow pier. The width of the pier is 324 mm and the height is 992 mm. The calculation is based on a plane stress situation. For the thickness a unit thickness of 1000 mm was chosen. To compare the numerical results with the results of a real pier with a thickness of, for instance, 100 mm, the force in a force-deformation diagram of a numerical simulation should therefore be divided by a factor of 10. The upper and lower layer and the little triangle in the top left side in Figure 5.4a are not part of the pier, but serve for the application of the boundary conditions and the load. The bottom side of the model is rigidly supported both in horizontal and vertical direction. After the dead weight of the pier has been taken into account, the upper side of the model is rigidly supported in vertical direction. The vertical edges are free. The pier is successively loaded by its dead weight and by a horizontal displacement to the right of the little triangle in the top left side of the pier.

The pier has been modelled at a semi-detailed level. This implies that the joint is modelled as an interface with zero thickness, in analogy with the finite element modelling according to Figure 3.5c. In the model fictitious unit dimensions are then used, that are of the same size as the original dimensions plus the real joint thickness. The interface's stiffness is deduced from the stiffness of the real joint. In Figure 5.4b the units are divided into triangular zones. The interfaces between the units behave

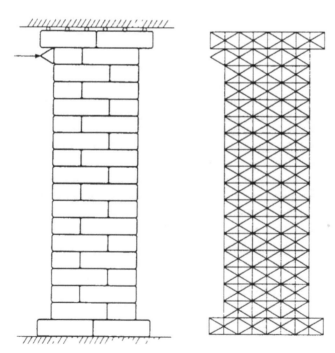

Figure 5.4. Modelling of the narrow pier with UDEC. a) Geometry with units/blocks, b) Division into zones.

non-linear elastic. When the bond tensile strength or shear strength is exceeded, the tensile strength and the cohesion are reduced to zero. This reduction occurs abruptly (elasto-brittle behaviour).

For the units in the piers an elastic-plastic Mohr-Coulomb model with tension cut-off was assumed. Here too, the stress is immediately reduced to zero (for the direction in which the tension occurs) after the tensile strength has been reached. The upper and lower layer and the little triangle on the top left side that are not part of the pier, are assumed to be linear-elastic. Also the vertical interfaces between the units have been kept linear-elastic, so that they cannot collapse.

Mohr-Coulomb's continuum model for the units is described by two parameters c and ϕ. Table 5.1 also includes the compressive strength f_c' or the units. The compressive strength is not a separate parameter but is related to c and ϕ according to: $c = f_c'$ $(1 - \sin \phi)/(2 \cos \phi)$.

5.2.2 *Results*

In Figures 5.5 and 5.6 typical results at the time of collapse of the pier are shown. The horizontal cracks between the pier and respectively the bottom layer and top layer (bottom left and top right) already occur under a small horizontal displacement of the upper side of the pier. After that the stiffness of the pier hardly changes. The force-deformation diagram shows a linear behaviour up to approximately 85% of the failure load. When the 85% of the failure load is exceeded, the deformation increases relatively more. After the top in the diagram has been reached, the load gradually reduces. Considering the force-deformation diagram, the pier does not collapse in a brittle manner, but has a relatively large toughness at its disposal.

As the load increases, cracks occur in the horizontal and vertical interfaces in the bottom left corner and the top right corner of the pier. In these areas the largest local deformations occur. In the middle of the pier fractures only occur in the vertical interfaces due to vertical shear. The final deformation of the pier shows a rotation in

Table 5.1. Material parameters for analysis of narrow pier with UDEC.

Component	Parameter	Symbol	Value	Dimension
Units	Elastic modulus	E	6300	N/mm^2
	Poisson's ratio	ν	0.1	–
	Tensile strength	f_t	1.76	N/mm^2
	Compressive Strength	f_c^l	9.40	N/mm^2
	Cohesion	c	2.71	N/mm^2
	Angle of friction	ϕ	30°	–
	Mass	ρ	1500	kg/m^3
Interfaces	Normal stiffness	k_n	$\times 10^9$	Pa/m
	Shear stiffness	k_t	24.3×10^9	Pa/m
	Bond tensile strength	f_t	0.26	N/mm^2
	Cohesion	c	0.26	N/mm^2
	Angle of friction	ϕ	30°	–
	Angle of dilatancy	ψ	0°	–

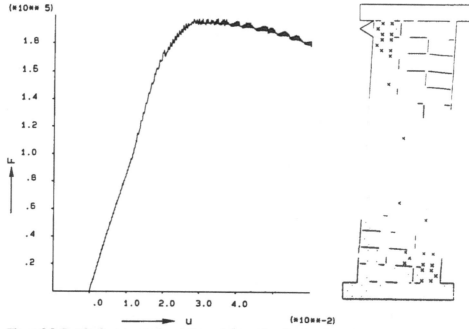

Figure 5.5. Results for narrow pier. a) Force-deformation diagram for point at top left corner where force is applied, b) Collapsed interfaces (stripes) and plastic points in the units (crosses).

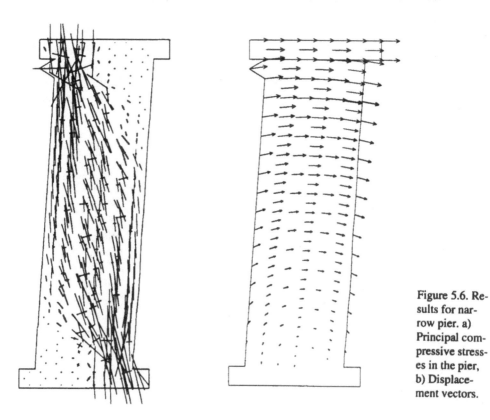

Figure 5.6. Results for narrow pier. a) Principal compressive stresses in the pier, b) Displacement vectors.

relation to the bottom right corner. This in contrast to the results of the study of the wide pier, which will be dealt with in the next section, in which under the same boundary conditions horizontal shear off takes place with stepwise cracking.

5.3 EVALUATION STUDY OF WIDE PIER WITH OPENING

5.3.1 *Modelling*

The calculations for the wide pier have been made in the context of the verification experiments for these piers that were executed within the framework of the B50 CUR committee.

Figure 5.7 shows the geometry of the pier. The width of the pier is 990 mm and its height is 1024 mm. Again based upon a plane stress situation with a unit thickness of 1000 mm. To compare the numerical results with the experimental results (thickness of wall 100 mm), the force in the load-deformation diagram should be divided by a factor of 10. The upper and lower layer and the little triangle on the top left side are not part of the pier, but serve for the application of the boundary conditions and the load. The bottom side of the model is rigidly supported in horizontal and vertical direction. The top side of the model, after the weight of the pier itself has been taken into account, is rigidly supported in vertical direction. The vertical edges are free. The pier is successively loaded by its dead weight and a horizontal displacement to the right of the little triangle on the top left side of the pier.

The (window)pier has been modelled at a semi-detailed level. In Figure 5.7b the units are divided into triangular zones. The interfaces between the units behave non-linear elastic. When the bond tensile strength or shear strength is exceeded, the tensile strength and the cohesion are reduced to zero. This reduction occurs abruptly (elasto-brittle behaviour). For the units of the pier an elasto-plastic Mohr-Coulomb model with tension cut-off was assumed. The upper and lower layer and the little tri-

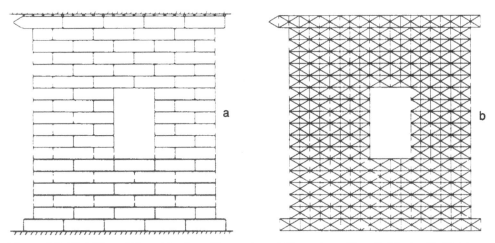

Figure 5.7. Modelling of the wide pier with opening. a) Geometry with units/blocks, b) Division into zones.

angle on the top left side which are not part of the pier are assumed to be linear-elastic. Also the vertical interfaces between these units have been kept linear-elastic, so that they cannot collapse. The Vijf Eiken unit is still the type of unit that was used. The values of the parameters are temporary values as the exact values from the research of the B50 CUR committee were not yet available when the calculations were made.

5.3.2 *Results*

In the Figures 5.8, 5.9 and 5.10 the characteristic results of the wide pier are shown when it almost collapses. In the load-deformation diagram of Figure 5.8a it can

Table 5.2. Material parameters for analysis of wide pier with UDEC.

	Parameter	Symbol	Value	Dimension
Units	Elastic modulus	E	6050	N/mm^2
	Poisson's ratio	v	0.14	–
	Tensile strength	f_t	2.47	N/mm^2
	Compressive Strength	f_c^1	9.0	N/mm^2
	Cohesion	c	2.6	N/mm^2
	Angle of friction	ϕ	30°	–
	Unit mass	ρ	1500	kg/m^3
Interfaces	Normal stiffness	k_n	2258 × 10^9	Pa/m
	Shear stiffness	k_t	844 × 10^9	Pa/m
	Bond tensile strength	f_t	0.22	N/mm^2
	Cohesion	c	0.70	N/mm^2
	Angle of friction	ϕ	37°	–
	Angle of dilatancy	ψ	0°	–

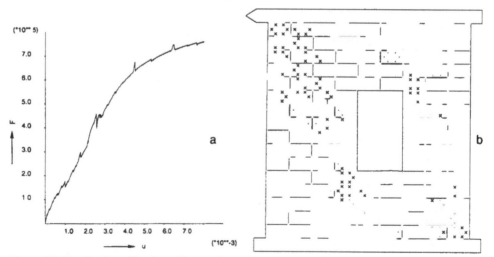

Figure 5.8. Results for wide pier with opening. a) Load-deformation diagram for point at top left where force is applied, b) Collapsed interfaces (stripes) and plastic points in the units (crosses).

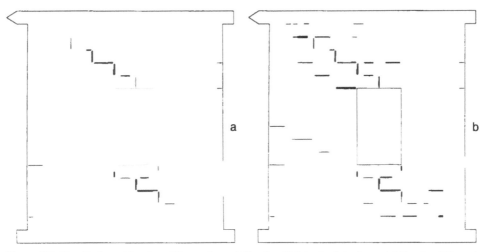

Figure 5.9. Results for wide pier with opening. a) Crack width in the interfaces, b) Crack width and shear slip in the interfaces.

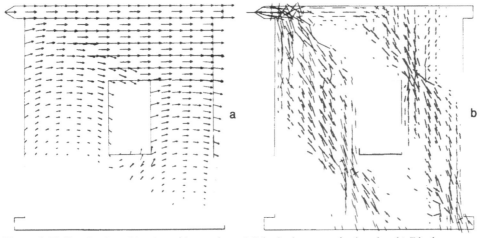

Figure 5.10. Results for wide pier with opening. a) Principal stresses in the pier, b) Displacement vectors.

clearly be observed that the top in the diagram has more or less been reached. In Figure 5.8b the collapsed interfaces are shown as well as the plastic behaviour in the units. In Figure 5.9a and 5.9b the interfaces are shown in which the crack openings and the shear slip are concentrated. In Figure 5.10a and 5.10b respectively the principal stresses and the displacement vectors are shown. The maximum principal compressive stress with a value of 23.7 N/mm² does not occur in the pier but in the upper edge where the displacement is applied. The maximum principal compressive stress in the pier amounts to 9.0 N/mm².

With regard to the force-deformation diagram, the expectation is that with an increasing displacement the force will gradually decrease as a result of which it shows a rather tough behaviour. The little peaks in the curve are moments at which a new crack occurs or plastic behaviour takes place in the unit. The deformation is clearly concentrated in a number of interfaces. Other interfaces that collapsed due to the exceeding of the tensile or shear strength, again transfer normal compressive stresses and shear stresses. Crack formation in the pier occurs diagonally in the direction of the upper and lower side of the opening. When looking at the stresses and the displacement vectors, the pier starts to behave as two separate piers.

In spite of the fact that the values of the parameters in the numerical model do not exactly coincide with the experimentally determined values, there is a great similarity in the crack behaviour and the load-deformation diagram between the theoretical model and the experiment itself.

5.4 COMPARISON OF DISTINCT ELEMENT METHOD WITH FINITE ELEMENT METHOD

The choice between the finite element method within DIANA and the distinct element method within UDEC depends on many factors. With the finite element method an iterative process is needed within each load or displacement increment to obtain a static equilibrium situation. During the iteration process a new tangent-stiffness matrix for the structure is set up and inverted once or several times. This is called an implicit method. The processes are time intensive, although the unit/block super elements considerably accelerate the process (refer to Section 3.4).

In the case of the distinct element method within UDEC it is not necessary to put up a system stiffness matrix. This is an explicit method with time increments, in which the elements are treated sequentially instead of simultaneously. This calculation process does also consume a considerable amount of time. Depending on the specific structure a great number of small time increments need to be carried out, especially when there is a considerable difference between stiff and weak parts.

With regard to the calculation time, the two methods will not differ that much. As for the results, both methods are able to simulate the diagonal cracking and the failure behaviour of transverse force piers. An advantage of the DIANA calculations is that softening can be taken into account, which is more realistic and offers an opportunity to predict snap-back behaviour. The choice between the two methods depends on these and many other factors, among which the availability of the program's source-code within the Netherlands, the possibilities to extend to three-dimensional and time-dependent effects, etc. [4].

CHAPTER 6

Case study cracking behaviour of walls under restrained shrinkage

6.1 INTRODUCTION

Crack formation due to restrained shrinkage is undoubtedly number one in the damage top-ten of the Dutch construction industry [62]. Unit-like materials such as calcium silicate unit, clay unit and concrete are characterised by a relatively high measure of shrinkage in combination with a relatively low tensile strength. As soon as the shrinkage or thermal cooling down of these materials is restrained by adjacent structure parts, tensile stresses occur that give rise to crack formation. This type of crack formation is, from a structural point of view, not that dramatic; the load bearing capacity of the structure is usually retained. From an aesthetic and psychological point of view, however, it is unacceptable. The cracks make beautiful facades look ugly and make people feel unsafe.

Throughout the years, constructors have come to terms with this phenomenon. Movement joints, alternatively called expansion joints or dilatation joints, are applied, allowing some mobility so that the wall, in spite of obstructions from floors and foundations can be subjected to a certain amount of shrinkage without cracking. However, if the movement joint scheme is either incorrect or the movement joint distance is too large, then crack formation will occur and the wall will create its own movement joint. A careful approach with small movement joint distances (a lot of joints) is not desirable either because of the necessary maintenance and the disturbing influence on the architectonic concept. Movement joints on or near corners are even experienced as extremely ugly, as can be seen from the following statement by a well-known German professor of architecture: *'Der Eckverband ist ein wesentliches Stilelement des Mauerwerkes. Mit der Anordnung solcher Eckfugen wird der Mauerwerksbau im wahrsten Sinne des Wortes 'um die Ecke gebracht'* (The corner forms an essential style element of the wall. The application of such joints on the corners means literally 'the end of the wall').

The above problem has led to this case study, which was carried out in close consultation with RCK and KNB. Because of the large number of influencing parameters [53, 63] modelling of crack behaviour due to shrinkage is not that easy. Based on previous experimental and analytical work the cracking process is only partly understood. In the building practice rough rules of thumb are used, based upon experience. In [64] it is stated that: 'the design of movement joints is based more on art than on science'. As a consequence, damage is insufficiently prevented, while the application of new masonry materials, for example glued masonry or masonry without vertical joints, or new types of structures, facades with different openings, for example, quickly give rise to problems. This chapter approaches the issue from a numerical point of view. The results from the previous chapters are utilised. It is demonstrated that simulations considerably increase the understanding and add to the development of rational calculation rules. Furthermore, a selection of results is shown. For background and details refer to [41].

6.2 DEFINITION OF THE PROBLEMS

6.2.1 *Behaviour of walls with obstructed shrinkage*

Figure 6.1 shows a wall on a foundation beam or part of a floor. Due to drying out and/or thermal cooling down the wall will shrink. We assume that the foundation

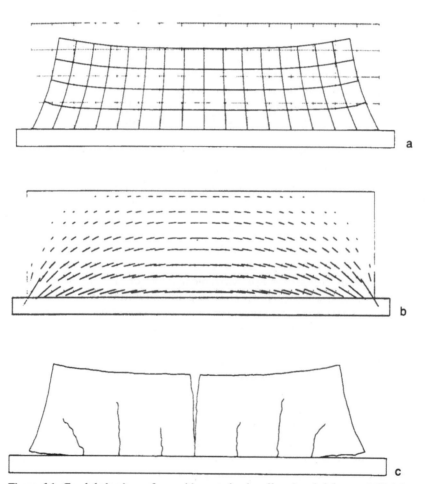

Figure 6.1. Crack behaviour of one-side restrained wall under shrinkage. a) Shrinkage of wall on restrained foundation beam or floor, b) Trajectories of the principal tensile stresses in the wall, c) Diagrammatical representation of crack behaviour.

will not shrink or to a far lesser extent than the wall. This is a realistic assumption because the foundation is subject to fluctuations of temperature to a far lesser degree, while possible hygroscopical shrinkage took place in previous phases of the construction process. For this reason the foundation will restrain or obstruct the shrinkage of the wall. With regard to the situation of Figure 6.1, the term one-sided restrained wall is appropriate. When an upper floor and/or side piers are present, than the wall will be more sided restrained. If no restraint at all is present, shrinkage can occur freely. The wall deforms without stress and will not crack. When restrained, on the other hand, tensile stresses will occur in the wall that will give rise to crack formation. Figure 6.1b gives an impression of the magnitude and the direction of the principal tensile stresses in the wall. We can observe a stress introduction zone near the ends and an almost undisturbed tensile stress field near the middle. The tensile stresses in the introduction zone could give rise to the creation of a horizontal crack at the boundary between wall and foundation, depending on the bond strength that is present there. As soon as the tensile stresses in the middle of the wall exceed the tensile strength, a vertical crack will appear. The location of the vertical crack will depend on the local scatter of wall strength. The 'weakest spot' near the middle will crack. Whether or not a vertical crack will appear, determines the need for movement joints for the given wall length. Figure 6.1c illustrates the crack pattern. Besides the mentioned primary crack near the middle, secondary cracks will occur.

6.2.2 *Influencing parameters*

The problem focuses on the fact that the above crack behaviour is dependent on a multitude of influencing parameters [56, 63], which can be divided into three categories:

1. *Boundary conditions.* The construction parts that surround the shrinking wall determine the extent of restraint as well as the occurring stress level and crack behaviour of the wall. These boundary conditions for the wall are caused by the entire construction of the building, including foundation beam, bedding or foundation piles, floors, storeys, side walls, etc. Aspects that play a role are the stiffness of the restraining construction parts, the bending stiffness of the restraining construction parts, the presence of one-sided versus multilateral restraint, the bond between wall and restraining construction parts, and the presence of a top load.

2. *Geometry.* Crack formation in structures of unit-like materials is accompanied by a size-effect and a shape-effect, refer, for instance, to [65]. A structure that is twice as large will not necessarily have a failure load or resistance against crack propagation that are simply twice as large. A similar argument applies to a change in shape. Examples of geometrical influencing factors, therefore, are the length of the wall, the height of the wall, the length/height ratio of the wall and the presence of openings. The influence of the length of the wall is, of course, directly connected to the issue of necessary movement joint spacings.

3. *Material properties.* Crack formation and crack propagation in unit-like materials require an elastic-softening model, as described in previous chapters. With such a model, the size-, shape- and boundary-effects can for the greater part be understood, refer for instance, to [66]. Elastic stiffness, tensile strength and tension-softening are the dominant material parameters. These properties of the wall depend

on the underlying properties of units and joints and the stacking structure, as examined in Section 4.3. The elastic modulus E of the wall determines the occurring stress level with restrained shrinkage. In case of complete restraint the following applies:

$$\sigma = \alpha_t \Delta T E \tag{6.1}$$

for thermal shrinkage and

$$\sigma = \varepsilon_h E \tag{6.2}$$

for hygroscopical shrinkage. Here, α_t is the thermal coefficient of expansion, ΔT is the decrease in temperature, and ε_h is the extent of hygroscopical shrinkage. Shrinkage restraint of a stiff wall (high E), therefore, leads to higher stresses and a larger risk of crack formation than with shrinkage restraint of a weak wall. The tensile strength determines the moment of crack initiation and the softening-part of the curve (fracture energy G_f^I) determines the extent of resistance against crack propagation.

6.2.3 *Present-day knowledge in the analytical and experimental field*

From a qualitative point of view, the effect of most of the influencing parameters is fairly understood. From a quantitative point of view little is known, which on the one hand is due to the complexity and interactions of the various influencing parameters, and on the other hand to the fact that crack propagation is a strictly non-linear phenomenon that is difficult to incorporate in a manual calculation approach. Nevertheless, analytical ways of approach have been developed which have resulted in formulae for the crack-free wall length as a function of some influencing parameters. The best-known for a one-sided obstructed wall, are:

$$\text{Copeland, 1957 [67]:} \quad L = -\ln\left(1 - \frac{f_t}{\varepsilon_k E R_a}\right)\frac{H}{0.23} \tag{6.3}$$

$$\text{Hageman, 1969 [68]:} \quad L = 3.52\frac{f_t H}{\varepsilon_k E} \tag{6.4}$$

$$\text{Schubert, 1983 [69]:} \quad L = 5\frac{f_t H}{\varepsilon_k E R_a} \tag{6.5}$$

In which L is the crack-free wall length, that is to say the wall length in which crack formation does not yet occur, f_t is the wall tensile strength, H is the wall height, ε_k is the total shrinkage (hygroscopical and thermal), E is the elastic modulus and R_a is the axial degree of restraint which has been defined as a ratio of the stiffness of the restraint in relation to the total stiffness of wall plus restraint:

$$R_a = \frac{E_b A_b}{E_b A_b + E_m A_m} \tag{6.6}$$

with subscript b for restraint (usually a concrete construction part), subscript m for masonry wall, subscript a for axial, and A for the cross-section The three formulae are diagrammatically shown in Figure 6.2. In practice a wall is dimensioned with a certain expected shrinkage which depends on the moisture content and the location of the wall in relation to the sun. The formula or the graph then gives the accompanying crack-free length which is equal to the necessary movement joint spacing.

All three formulae include the tensile strength in the numerator and the elastic modulus in the denominator. Therefore, an increase of the tensile strength will result in a greater crack-free wall length (read: admissible movement joint spacing), while an increase in the stiffness will result in the crack-free wall length becoming smaller, which is plausible.

The three formulae also show that a shorter wall can withstand a higher shrinkage without crack formation in comparison to a longer one. The relation is strictly non-linear and asymptotic. The vertical asymptote reflects that all relatively long walls crack at almost the same shrinkage. With long walls the introduction zone effect of the stresses (Fig. 6.1b) has been dampened at the location of the midsection, so that the wall in the middle is 'not conscious of what happens on both ends'. The tensile stress distribution in the midsection is uniform and the crack will, as soon as it originates, immediately be present along the entire height. With relatively short walls the stress distribution in the midsection is non-uniform. The crack will than start at the bottom of the wall and will slowly propagate towards the top. Redistribution is possible. Short walls have an important 'reserve' before they completely crack right through, which is expressed by the strong increase of the admissible shrinkage. These essential aspects have been illustrated in the Appendix.

Although the three formulae show roughly the same tendency, there is a considerable scatter. There is lack of understanding [56]. One of the reasons being that the

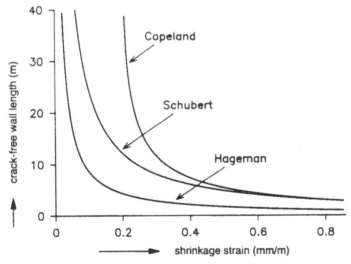

Figure 6.2. Crack-free wall length L as function of imposed shrinkage E_k. Diagrammatical reproduction of Equations (6.3), (6.4) and (6.5) according to respectively Copeland, Hageman and Schubert. Valid for one-side obstructed wall when $f_t = 0.5$ N/mm^2, $E = 5000$ N/mm^2, $H = 2400$ mm, $R_a = 0.5$.

formulae are based on analytical solutions [70] and photo-elastic research [71, 67] under the assumption of linear-elastic material behaviour. A crack propagation theory with softening was not available in this computerless era. Whether or not cracks occurred, was roughly determined on the basis of the 'average tensile stress in the wall'. Some of the influencing factors mentioned in the previous section were not or not sufficiently taken into account. Moreover, there are different assumptions concerning the way in which the bond between wall and restraining construction part should be taken into account. It is illustrative, for example, that an empirical reduction factor for the degree of restraint R_a has been introduced in subsequent publications [63], so that the definition according to Equation (6.6) is not uniformly maintained. Another source of misunderstandings is the lack of a suitable definition for the concept of 'crack'. How wide does a crack need to be for the various experts to speak about a visible crack?

The analytical formulae have partially been tested by means of practical experience and experimental laboratory research. Practical experience has led to correction factors and coefficients which give the formulae a clearly empirical character. An extensive experimental research has taken place in the Netherlands at the Research Centre for the Calcium Silicate Unit Industry (*Researchcentrum voor de Kalkzand-steenindustrie*) (Berkers & Rademakers [56]). This 'long wall research' has considerably increased the understanding of this issue, but has not yet been able to provide a coherent answer to the effect of the various influencing parameters. It was concluded that, therefore, additional research was necessary in the form of, for example, numerical calculations.

6.3 NUMERICAL RESEARCH

Preliminary calculations [41] have led to the setting up of a simple but insight-providing model with one vertical potential crack in the middle of the wall. In this section this modelling is worked out for a wall length of 6 m. In the sections that follow variation studies will be carried out. The chapter is concluded with a feed back to the above calculation rules and with the practical relevance.

6.3.1 *Modelling*

We will examine a one-sided restrained wall with a length of 6m, a height of 2.4m and a thickness of 100 mm. Because of the symmetry only one half of the wall needs to be considered. The wall is standing on a concrete foundation beam (or part of a floor) with dimensions of 200×200 mm. The foundation beam determines the extent of restraint. Initially, the foundation beam was restricted in vertical direction, so that bending and curling up are completely prevented ($EI_{\text{restraint}} = \infty$). In horizontal direction both the wall and the beam can move freely. Horizontal supports have only been located along the axis of symmetry of the beam.

The joint between wall and foundation beam has been modelled with interface elements. In [41] it has been shown that a horizontal crack in this joint can make the crack-free wall length smaller, depending on the strength and stiffness properties of the joint. In this section a conservative approach was followed so that this positive

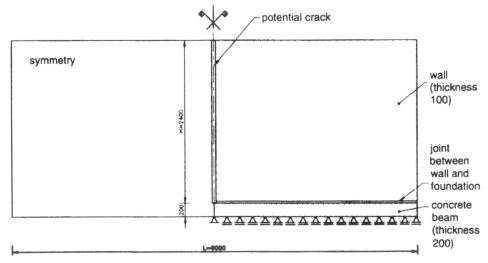

Figure 6.3. Modelling of half a wall (symmetry) on bending stiff foundation beam. In the middle a vertical, potential tension crack has been applied.

effect has not yet been taken into account. The behaviour of the joint between wall and foundation beam is assumed to be linear-elastic.

In the middle of the wall interface elements have been placed as a vertical, potential crack. Initially, these elements have a high dummy stiffness, so that they are not subject to deformation. When the tensile strength is exceeded, the softening model starts to work and the crack will gradually open up.

Table 6.1 contains the chosen material parameters. The values do not specifically correspond with a certain type of wall that consists of clay units or calcium silicate units, but form a sort of average that approaches some of the types of masonry consisting of both clay units and of calcium silicate units. From the chosen elastic moduli E and the cross-sections A of both masonry wall and concrete beam, it can be concluded that the axial degree of restraint R_a according to Equation (6.6) is in this example equal to 0.5.

With the tensile strength and tensile softening parameters it should be realised that these apply to the imaginary straight crack. From the detailed analyses of parts of wall (refer to Section 4.3) it is known that, depending on the underlying properties and stacking structure, a stepwise crack will usually appear with a saw tooth-like stress-crack width diagram. The direct modelling of such a stepwise crack in the undamaged walls was, for the time being, considered too detailed. Instead, an imaginary straight crack was assumed with an increased value of the fracture energy ($G_f^I = 0.05$ J/m^2) in order to take into account the extra toughness due to the stepwise shape [41].

Initially, attempts were made to include the time-dependant behaviour in the research in the form of non-linear shrinkage and creep [41]. In reality a lot of time-dependent effects occur simultaneously, among which hygroscopical shrinkage, temperature effects, creep, time-dependent development of glue/mortar strength, carbonation, etc. Addition of shrinkage and creep is, therefore, still an incomplete ap-

Table 6.1. Material parameters for the analysis of shrinkage of a 6 m-long wall on a foundation beam ($R_a = 0.5$) with restraint bending.

Component	Parameter	Symbol	Value	Dimension
Masonry	Elastic modulus	E	5000	N/mm²
	Poisson's ratio	v	0.2	–
	Mass	ρ	1800	kg/m³
Concrete beam	Elastic modulus	E	30000	N/mm²
	Poisson's ratio	v	0.2	–
	Unit mass	ρ	2400	kg/m³
Joint between wall and	Normal stiffness	k_n	333	N/mm²
foundation	Shear stiffness	k_t	139	N/mm²
Potential crack	Normal stiffness	k_n	10⁶	N/mm²
	Shear stiffness	k_t	10⁶	N/mm²
	Tensile strength	f_t	0.5	N/mm²
	Mode-I fracture energy	G_f^I	0.05	J/m²

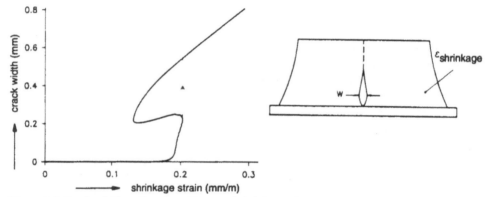

Figure 6.4. Crack width as function of imposed shrinkage. Wall length 6 m, $R_a = 0.5$, $EI_{beam} = \infty$. The moment of snap-through of the crack to the upper plane has been shown diagrammatically.

proach and for the time being appeared to generate more new questions than anwers. Finally it was decided, in analogy with the analytical considerations [67-69], to start with simple time-independent material behaviour. If desired, the influence of creep can roughly be estimated by reducing the elastic modulus.

6.3.2 *Results*

In the calculations the dead weight of wall and foundation beam was first applied. Subsequently, a linear time-independent shrinkage was imposed on the wall (initial strain). This shrinkage was gradually increased, in small steps or increments. Figure 6.4 shows the calculated development between the imposed shrinkage and the calculated maximum crack width. The maximum crack width can be found somewhere along the vertical midsection, usually at the bottom of the wall, just above the foundation. The location of the maximum crack width can vary from increment to increment. In the final stage, when the crack had propagated to the upper plane, the maximum crack width occurred at the top. The crack opening profiles can be recognised in Figure 6.5 where the deformed element mesh has been illustrated for various stages of the shrinkage process. Figure 6.6 shows the accompanying principal stresses.

As long as all the stresses in the potential crack are lower than the tensile strength, the system behaves linear-elastic. In the diagram we can see that this linear-elastic stage remains unaltered up to a shrinkage of about 0.15 mm/m. Until that level of shrinkage the crack width is zero. At that moment the tensile strength is reached at the bottom of the wall and the potential crack starts to open up. From that moment on, the diagram shows an increase in the crack width with increasing shrinkage. From the deformations (Fig. 6.5) it appears that the crack gradually propagates upwards and opens up further and further.

With a shrinkage of about 0.20 mm/m and a crack width of about 0.24 mm a strange 'arc' starts to develop in the diagram. Instead of increasing, the shrinkage temporarily decreases (swelling). Also the calculated crack width shows a temporary decrease. At first sight this phenomenon seems remarkable. It is known as snap-back behaviour and it corresponds with reflective cracking towards the upper plane. Figure 6.5c and 6.5d give a clear illustration. These figures show the deformed element meshes respectively just before and just after reflective cracking takes place (with position (c) and (d) in the diagram, as indicated). In Figure 6.5c the crack is still incomplete and stretches across about three-quarters of the wall height. In Figure 6.5d the crack is present across the full height. With monotonously increasing shrinkage this reflective cracking will occur suddenly, as indicated by means of the dotted line in Figure 6.4. The snap-back phenomenon has already been discussed with the detailed studies in Section 4.3.4 and now also appears to occur at structural level.

From practical experience and experimental research it is known that crack formation in long walls is accompanied by a loud bang. Reflective cracking is an explosive phenomenon. This entirely corresponds with the process as described above. Usually, divergence would have occurred in the numerical analyses due to unstable crack propagation. In this project a special arc-length procedure was used (refer to Section 3.2.2), which is able to keep the propagation of the crack numerically stable, by making the wall temporarily swell instead of shrink. In that case a dynamic jump is not found, but a quasi-static snap-back. An advantage of this method is that the failure behaviour and the post-cracking behaviour are made comprehensible. Besides, quantitative information is obtained with regard to the crack width. The maximum crack width increases from 0.24 mm before reflective cracking to 0.55 mm immediately after. The maximum crack width can be found in the top of the wall (Fig. 6.5d). An interesting detail is that in practice there exists a widespread misunderstanding that the crack starts at the top. Figure 6.5d (and 6.1c) suggest that as well, but this is a false suggestion. The crack starts at the bottom of the wall.

From the principal stress trajectories (Fig. 6.6) one can deduce the gradual softening at the location of the crack. The tension band, which at the beginning fans out widely across the entire midsection, concentrates with increasing crack formation near the bottom side and the upper side of the wall without cracks. After reflective cracking it is impossible to transfer any further tension to the upper side. The wall

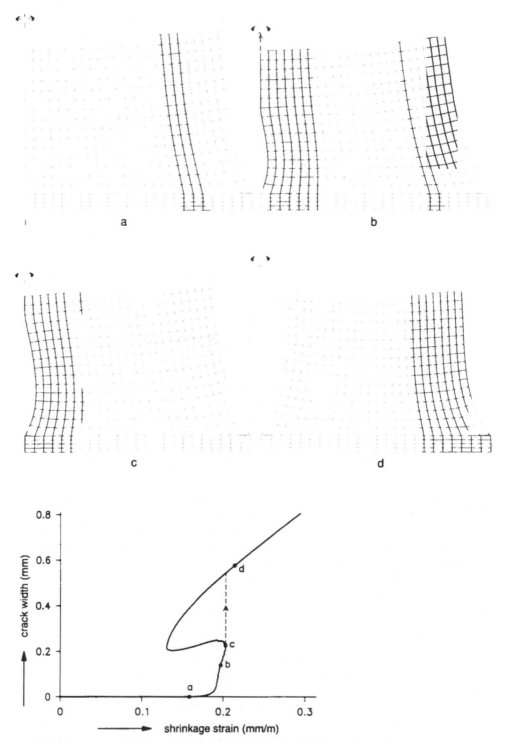

Figure 6.5. Deformations at increasing stages of the loading process (enlargement factor 1200). Wall length 6 m, $R_a = 0.5$, $EI_{beam} = \infty$.

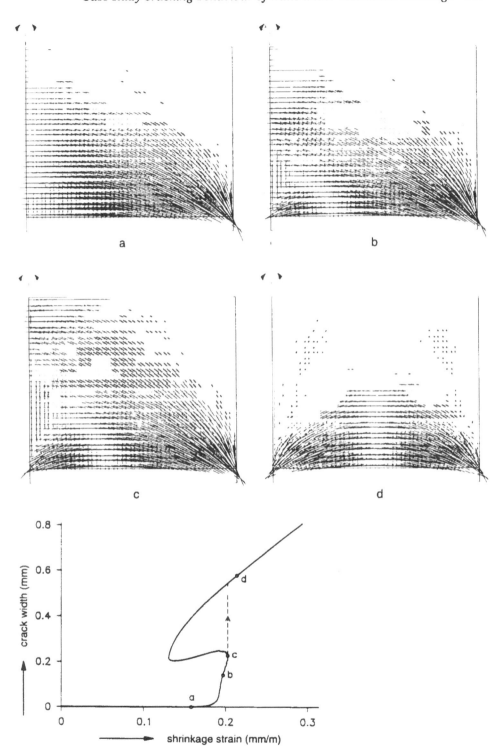

Figure 6.6. Principal stresses at increasing stages of the loading process. Wall length 6 m, R_a = 0.5, $EI_{beam} = \infty$.

has then split up into two parts. What remains is a tension band in half the wall, in analogy with the original tension band in the entire wall. In the calculation as shown here, the effect of the new tension band has not been limited. If a potential crack would have been imposed at a quarter of the wall length, then the process would repeat itself and the wall would split up into four parts [41].

The major conclusion from this section is that it is possible to predict the complete redistribution process from crack initiation, crack propagation up to and including snap- through and reflective cracking. The behaviour that has been found is plausible and results in practically relevant, quantitative information with regard to crack width and crack height with increasing shrinkage. This is a step forward when compared to the analytical formulae in which global notions were made use of, such as the 'average tensile stress in the wall' and a quantitatively unclear definition of the concept of 'crack'.

6.4 INFLUENCE OF BOUNDARY CONDITIONS

6.4.1 *Axial degree of restraint*

The calculation that was departed from Section 6.3 referred to an axial degree of restraint R_a of 0.5. This degree represents the ratio between the stiffness of the restraining beam and the total stiffness of beam plus wall, according to Equation (6.6). The degree of restraint is a relative stiffness ratio which lies between 0 (unrestrained, $E_bA_b \ll E_mA_m$) and 1 (completely restrained, $E_bA_b \gg E_mA_m$).

Numerical variation studies have been carried out for $R_a = 0.25$, 0.75 and 1.0 instead of 0.5. All the other parameters and boundary conditions have been assumed equal to that of the reference calculation. Figure 6.7 shows the influence of R_a on the calculated shrinkage-crack width relationship.

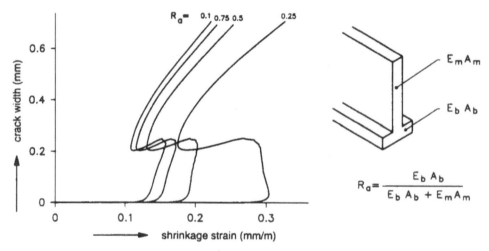

Figure 6.7. Influence of axial degree of restraint R_a on shrinkage-crack width diagram. Wall length 6 m, $EI_{beam} = \infty$, other parameters as in Section 6.3. The calculation with $R_a = 0.5$ corresponds with that of Figure 6.4.

The four curves have nearly the same shape. In all the cases the crack width increases up to a similar level of approximately 0.24 mm, after which reflective cracking occurs. The difference lies in the level of shrinkage that accompanies reflective cracking. This critical shrinkage value increases when R_a decreases. The diagram is subject to translation to the right when R_a decreases. This behaviour was to be expected. The lower the degree of restraint, the lower the tensile stresses and the longer crack formation is delayed, which has a positive effect. In case of the extreme situation with $R_a = 0$ no crack would appear at all since the wall can shrink freely. The above tendency has in a certain way been included in the existing formulae of Copeland (Eq. 6.3) and Schubert (Eq. 6.5). In both the formulae R_a and ε_k occur in the form of a product (ε_k is interpreted as the critical shrinkage belonging to reflective cracking for the given wall length). If the other parameters (wall length, wall height, tensile strength and elastic modulus) remain constant, than the formulae show a halving of ε_k when R_a is doubled; the product of the two remains constant, or in other words, the parameters are inversely proportional. When with $R_a = 0.25$ a certain value of ε_k is found, than $R_a = 0.5$ leads to $1/2\varepsilon_k$, $R_a = 0.75$ leads to $1/3\varepsilon_k$, and $R_a = 1.0$ leads to $1/4\varepsilon_k$. The numerical values for ε_k with reflective cracking can be measured in Figure 6.7 as the shrinkage values whereby the curves have a vertical tangent line. This results in $\varepsilon_k = 0.304$ mm for $R_a = 0.25$, $\varepsilon_k = 0.202$ mm for $R_a = 0.5$, $\varepsilon_k = 0.172$ mm for $R_a = 0.75$ and $\varepsilon_k = 0.158$ mm for $R_a = 1.0$. Although this result is not inversely proportional, the tendency, however, for an ε_k to decreases less and less when R_a becomes larger and larger, is fairly similar to that of the existing formulae.

A possible explanation for this similarity is that the redistribution route to reflective cracking is not subject to any drastic changes with variation in R_a. The linear-elastic observations which are the basis for the analytical formulae appear in this case to work rather favourably. It should be noted, however, that in this section only one combination of f_t, E, L and H is observed. Additional variation studies are necessary before it will be known whether the existing formulae with regard to this point are confirmed or should be adapted. In later publications [63], Schubert states that R_a in practice is situated somewhere between 0.4 and 0.8. However, it is not fully clear whether he maintains the same definition for R_a as in Equation (6.6), or whether some reduction factor has been added for e.g. the connection between wall and restraint.

6.4.2 *Bending stiffness of the restraint*

In the previous section the bending stiffness of the restraining concrete beam was held constant ($E_b I_b = \infty$), while the axial stiffness $E_b A_b$ varied. In this section the axial stiffness is held constant and the bending stiffness varies. The total cross section of the concrete beam (or part of a floor) is held constant, but the width-height ratio was adapted. Two ratios were investigated: $w \times h = 1800 \times 200$ mm and $w \times h = 450 \times 800$ mm. In both cases the total cross section is 0.36 m^2, so that the axial degree of obstruction R_a is equal to 0.9. If we define, in analogy with the axial degree of obstruction R_a, a bending-obstruction degree R_i according to:

$$R_i = \frac{E_b I_b}{E_b I_b + E_m I_m} \tag{6.7}$$

than for the first variant $R_i = 0.06$ and for the second $R_i = 0.5$. The value of this degree is limited. Axial stiffnesses may be added to each other, but bending stiffnesses obviously not. Actually, it is the bending stiffness of wall and beam as a composite girder which is of importance. Nevertheless the degree gives a fair indication.

The vertical bearings from Figure 6.3, which lead to $E_b i_b = \infty$ and $T_i = 1$, have in the two calculations been replaced by a bedding. The bedding constant k_n has been taken as 0.06 N/mm³. As soon as the beam curls up and separates from the underground, the bedding constant becomes zero. This no-tension behaviour of the bedding has been modelled by means of interface elements. The other material parameters and the dimensions remained unaltered, as in Section 6.3.1. An element mesh was used that is twice as course. This has no adverse consequences. In [41] an example is shown from which it appears that the calculation results are objective when the mesh becomes finer, owing to the use of tension softening in discrete cracks.

Figure 6.8 shows the influence of the bending stiffness on the calculated shrinkage-crack width diagram. With an increasing bending stiffness of the restraint, the shrinkage at which reflective cracking occurs, decreases. The influence is very large. Where the wall with $R_i = 1$ already cracks with a shrinkage of 0.16 mm/m, the wall with $R_i = 0.5$ does not crack until the shrinkage is 0.37 mm/m and the wall with $R_i = 0.06$ only cracks at a very late stage with a shrinkage of 0.85 mm/m. When the bending stiffness reduces, there is a considerably larger reserve present before reflective cracking occurs. The starting point of the three curves more or less coincide.

Figure 6.9 gives an impression of the deformed element meshes. Beam and wall act as a composite girder which bends due to unequal shrinkage. The ends curl up as a consequence of the no-tension bedding. In the composite girder there is a stress situation in the midsection which significantly differs from when $R_i = 1$. At the bottom, where the crack begins, again a tension zone can be found, but at the top of the wall there now is a compression zone present which prevents propagation of the crack to the top. The crack can hardly penetrate the compression zone. This explains

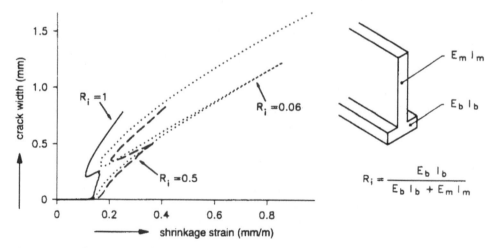

Figure 6.8. Influence of bending-restraint degree R_i on the shrinkage-crack width diagram. Wall length 6 m, $R_a = 0.9$, other parameters as in Section 6.3.

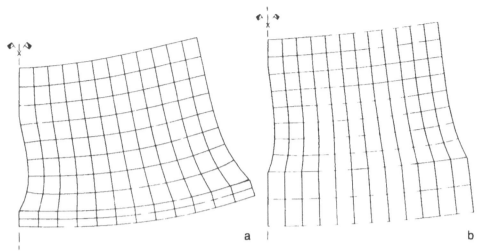

Figure 6.9. Deformed element networks when the bending-restraint degree R_i varies. a) $R_i = 0.06$, b) $R_i = 0.5$.

why there is a larger reserve before reflective cracking occurs. If we compare the results with that of the previous section, than it is striking that with a varying axial degree of restraint the shrinkage-crack width curves all have a nearly similar shape but occur in a different position, whereas with a varying bending stiffness the curves all begin at the same location but differ strongly in shape and length. Apparently R_a determines the crack initiation, whereas R_i mainly influences the crack propagation.

In analytical observations [63, 69] attention is paid to the influence of bending stiffness. Although not explicitly mentioned the analytical formula seem to depart from an infinitely bending stiff obstruction. Everything is explained from the axial degree of obstruction, supplemented with an empirical reduction [63]. The numerical results give rise to a reconsideration of this method. From a practical point of view, the idea of a 'complete' degree of restraint as a figure between zero and one is simple and attractive. Subsequent studies will have to prove whether the combination of the axial and bending-obstruction degree can be achieved in one new degree, or whether the underlying influencing factors should be taken into account separately.

Assessment of the bending stiffness of the obstruction is not so easy in practical situations. Foundation beam, bedding and possible piles play a role. In the case of buildings with several storeys made of load-bearing masonry, beams and floors cannot freely curl up. Vertical displacements are partially hindered. This would be an argument in favour of the assumption of a high bending stiffness of the restraint, which according to the calculations has an unfavourable effect on the cracking behaviour. For practical purposes, it should also be realised that not exclusively the moment of reflective cracking can be critical, but also the crack width requirement of the still incomplete crack. From Figure 6.8 it can be concluded that for relatively bending weak obstructions the crack width (at bottom of wall) strongly increases before reflective cracking occurs. For $R_i = 0.06$ the predicted crack width that precedes reflective cracking is already 1.2 mm, which is far from acceptable.

6.4.3 *Two-sided restraint*

So far, a situation has been considered in which the top edge of the wall can deform freely. In practice, an restraining floor or beam can often be found above the wall, especially when it concerns a loas-bearing wall. This is referred to as a two-sided obstructed wall. We will observe the same modelling and parameters as with the calculation we departed from Section 6.3, complemented with a part of a floor on top. The stiffness and bending stiffness of the part of the floor on top have been assumed similar to that of the part of the floor laying underneath ($R_a = 0.5$, $R_i = 1$). Figure 6.10 shows the influence on the shrinkage-crack width diagram.

The influence is considerable. With two-sided restraint the crack snap- through almost immediately after initiation. In both cases crack initiation occurs in combination with a shrinkage of approximately 0.15 mm/m. Whereas it was possible to subsequently increase the shrinkage to 0.20 mm/n with one-sided restraint, reflective cracking already occurs with two-sided obstruction with a shrinkage of approximately 0.155 mm/m. The reserve with redistribution of tensile stresses and gradual crack propagation has nearly disappeared. Figure 6.11 shows deformations. The elastic stress distribution at both the bottom and the top of the wall gives rise to crack formation. The two cracks propagate and meet each other soon [41], which is accompanied by snap-back behaviour and a transfer to another equilibrium system. The final situation shows an extended crack with the maximum crack width in the middle of the wall.

There are already existing methods to take the influence of two-sided restraint into account. In Copeland's logarithmic function (Eq. 6.3), which has been adopted later on by others as well, the constant 0.23 is doubled to 0.46 for two-sided obstruction [55]. With the parameters and wall dimensions that were chosen here, the critical shrinkage ε_k would then decrease from 0.455 for one-sided to 0.290 mm/m for two-sided obstruction, a reduction to 64%. The present calculation shows a reduction from 0.20 to 0.155 mm/m (78%). Here too, it must be mentioned that only one combination of parameters, boundary conditions and geometry was investigated. As the wall length and R_i increase, the difference between crack behaviour for both one-

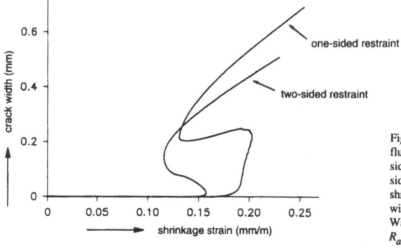

Figure 6.10. Influence of one-sided versus two-sided restraint on shrinkage-crack width diagram. Wall length 6 m, $R_a = 0.5$, $R_i = 1$.

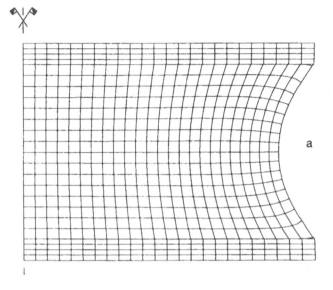

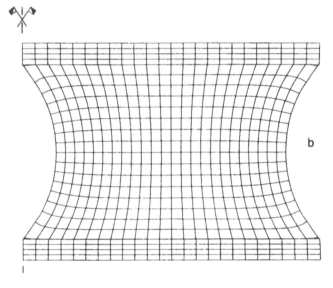

Figure 6.11. Deformations in the presence of two-sided restraint, a) Uncracked, linear-elastic stage, b) Final situation with extended crack.

sided and two-sided restraint becomes smaller [41]. In practical situations it is also possible that a three-sided or four-sided restraint occurs, depending on the presence of side piers and front facades, bending stiff or otherwise.

6.4.4 *Friction/slip and top load*

So far, the behaviour of the joint between wall and foundation beam has been assumed linear-elastic. Possible friction and slip have not been taken into account. The stresses in the joint between wall and foundation beam can, as a consequence, increase unlimited. In the stress diagrams of Figure 6.6 this can be recognised by the relatively large principal tensile stresses near the wall ends. This seems not very realistic when this joint has been constructed with a relatively weak mortar or even (consciously) with sliding foil.

Additional calculations have been carried out in which friction/slip along the joint between wall and foundation was permitted. The Coulomb friction model from Figure 3.9 was used with cohesion $c_u = 0.75$ N/mm^2, Mode-II fracture energy $G_f^{II} = 0.1$ J/m^2, angle of friction tan $\phi = 0.75$ and angle of dilatancy tan $\psi = 0.2$. The other parameters, boundary conditions and geometry are similar to the calculations that were departed from Section 6.3 ($R_a = 0.5$, $R_i = 1$). Since the level of the friction forces depends on the vertical prestress, the top load on the wall has been varied as well. In the calculation that was departed from there was no top load present; only the wall's dead weight caused a limited prestress $\sigma_{w-f\,joint}$ on the joint between wall and foundation. Three calculations with friction have been carried out, for:

– Top load zero, total load 1× dead weight, $\sigma_{w-f\,joint} = 0.043$ N/mm^2;
– Top load 4× dead weight, total load 5× dead weight, $\sigma_{w-f\,joint} = 0.215$ N/mm^2;
– Top load 9× dead weight, total load 10× dead weight, $\sigma_{w-f\,joint} = 0.43$ N/mm^2.

Figure 6.12 shows the influence on the shrinkage-crack width diagram. Firstly, we compare the two calculations without top load. This appears to be the two extremes. The difference is extremely large. In the calculation with slip there occurs no reflective cracking at all. The shrinkage can be increased to a very high level, while the crack width only increases gradually. The graph has been cut off at a shrinkage of 0.8 mm/m and a crack width of 0.32 mm, but it was possible to continue the calculation up to a shrinkage of 1.8 mm/m without any significant change. This is caused by the fact that the joint between wall and foundation beam already cracks over a great length at an early stage. At a shrinkage of 0.15 mm/m the joint between wall and foundation beam had already cracked over half its length and at a shrinkage of 0.30 mm/m already over three-quarter of its length. The adhesion between wall and beam has then been broken. Transfer of force no longer takes place (due to cohesion-softening shear stress and normal stress along the joint between wall and foundation beam gradually decrease to zero; when there is no top load there remains a zero stress situation near the apex of the residual Coulomb friction criterion in Fig. 3.9).

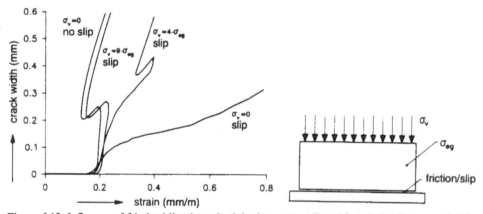

Figure 6.12. Influence of friction/slip along the joint between wall and foundation beam on shrinkage-crack width diagram. Calculations with and without top load. Wall length 6 m, $R_a = 0.5$, $R_i = 1$. The result without slip and with $\sigma_v = 0$ corresponds with Figure 6.4.

The wall can freely shrink at both ends. This restricts the tensile stress from being built up in the midsection and, as a consequence, the propagation and opening up of the vertical crack. In Figure 6.13 this behaviour has been illustrated with stress and deformation diagrams. It can clearly be observed that the part of the wall above the cracked joint between wall and foundation beam is an almost 'dead' part which can freely shrink.

When top load is added, the compressive stress on the joint between wall and foundation beam increases, and a residual shear stress can be transferred. The behaviour then lies somewhere between the two extremes. For the two levels of top load that were observed, an unlimited increase of shrinkage without reflective cracking appears to be no longer possible. Reflective cracking occurs with a top load of 4 and 9 times its own weight, when the shrinkage is respectively 0.39 mm/m and 0.23 mm/m. It is clear from Figure 6.12 that the shrinkage-crack width curve approaches the curve without friction/slip when the top load increases. With increasing top load the shear stress that is transferred along the joint between wall and foundation beam can reach such a high value that hardly any or no slip whatsoever occurs.

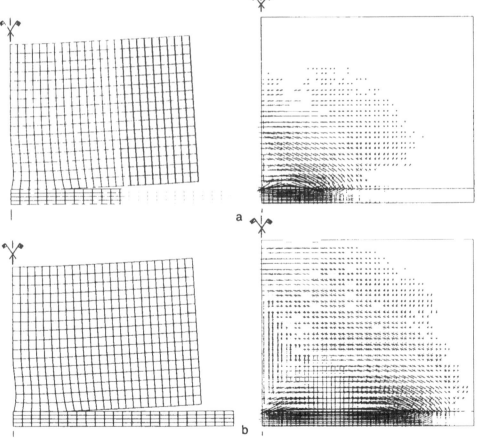

Figure 6.13. Deformations and principal stresses for calculations with friction/slip along the joint between wall and foundation. a) Without top load, with shrinkage 0.30 mm/m, b) With top load 9 times its dead weight, with shrinkage 0.23 mm/m.

The stress combination will stay within Coulomb's failure envelope. The calculation without slip can, therefore, be interpreted as a result for a wall with a very large top load.

The calculations offer a quantitative insight into the increase of crack sensitivity with an increasing top load. In the detailed studies of Section 4.3.9 it has been shown that top load also brings about a second effect, namely an increase of the masonry tensile strength. This effect works the opposite way (refer to Section 6.5), but has not yet been taken into account in these calculations. Its practical relevance lies in the difference between supporting and non-supporting walls. Load- bearing walls are more critical with regard to reflective cracking than non-supporting walls. Besides, supporting walls are two-sided restraint, so that another extra negative effect develops. Non-supporting walls are less critical and often unrestrained on top (PUR-joint). There may, however, be side-restraints or other restraints present. The angle of friction ϕ and cohesion c_u have not yet been varied. The tendency of the results, however, is an indication that it is advisable to place sliding foil at the wall-ends. This technique is sometimes used in the construction industry to allow movement of a wall to the maximum in relation to the surrounding construction parts. The reduction of friction and cohesion will have a positive effect. Other influences and correlations have been discussed in [41].

Finally, it should be noted that these calculations impose a high standard on the numerical procedures and material models. There are two active cracks, the vertical tension crack and the horizontal shear crack, which both simultaneously consume energy. Proper selection has to be made in the arc-length methods and in the size of the increments to reach convergence.

6.5 INFLUENCE OF MATERIAL PARAMETERS

A second category of influencing parameters as mentioned in Section 6.2.2 includes the material properties of the wall. Figure 6.14 shows the influence of the elastic modulus E, tensile strength f_t and fracture energy G_f^I. Point of departure was the reference case from Section 6.3. The parameters have both been doubled and halved in comparison to the values that were departed from $E = 5000$ N/mm^2, $f_t = 0.5$ N/mm^2 and $G_f^I = 0.05$ J/m^2. For each calculation only one parameter was changed; the others were kept constant. For the variation of E the thickness of the foundation beam was adapted in order to keep the axial degree of obstruction R_a according to Equation (6.6) constant. The most interesting aspect is the critical shrinkage ε_k at the moment of reflective cracking. For the referential calculation ε_k was equal to 0.20 mm/m.

The influence of the elastic modulus is the most pronounced. A doubling of E implies a stiffer wall. This results in an increase of stresses and crack sensitivity. The critical shrinkage lies already at 0.137 mm/m. On the other hand, a flexible wall with a halved E modulus has a positive effect. Then the critical shrinkage increases up to 0.344 mm/m. For the tensile strength the effect is of course the other way around. A doubling of f_t delays reflective cracking to a shrinkage of 0.324 mm/m, while halving of f_t leads to $\varepsilon_k = 0.174$ mm/m. Both the elastic modulus and the tensile strength influence the moment of crack initiation. The first bend away from the zero line in Figure 6.14 occurs at different levels of shrinkage. When the fracture energy is var-

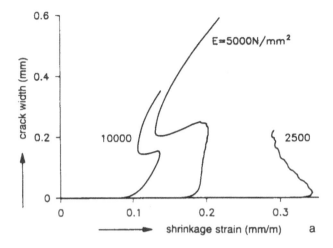

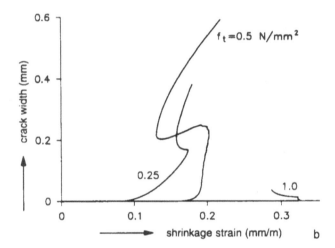

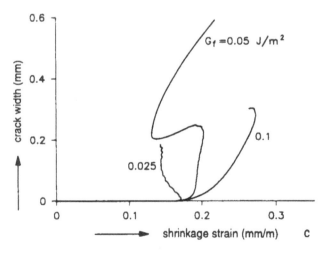

Figure 6.14. Influence of material properties on the shrinkage-crack width diagram. a) Elastic modulus, b) Tensile strength, c) Fracture energy.

ied this is not the case. The moment of crack initiation is not influenced, as opposed to the extent of reserve until reflective cracking occurs. The halved fracture energy leads to such a brittle behaviour that immediate reflective cracking occurs, at a shrinkage of 0.171 mm/m. The doubling leads to a tough behaviour with a reserve up to a shrinkage of 0.273 mm/m.

The existing design formulae (Eqs 6.3-6.5) all include the quotient $(f_t/E\varepsilon_k)$. This implies that f_t and ε_k are proportionate and that E and ε_k are inversely proportionate. The numerical results give rise to an adaptation of these formulae, that is to say for relatively short wall lengths as observed here (6 m). In the case of very long walls the redistribution route to reflective cracking is hardly relevant and the formulae that are based on linear-elastic behaviour are valid indeed. So far, fracture energy is lacking in the design formulae. The present results give rise to an introduction of this parameter. A promising approach is the use of a brittleness number that includes both f_t, E and G_f^I [35].

Sensitivity studies as described in this section, can add to a safe introduction of new materials or processing methods. Through experimental determination of the material parameters, if necessary, complemented by detailed analyses from Section 4.3, it is possible to determine tensile strength, elastic modulus and brittleness of the composite. The simulations that are shown here, consequently show how the modified properties behave at construction level. Examples of current questions with regard to material influences on crack behaviour are the (unfavourable) influence of the increased E and the (favourable) influence of the increased f_t for glued masonry made out of clay units (does the one compensate the other?), and the (favourable) influence of a decreased E for masonry without vertical joints.

6.6 INFLUENCE OF GEOMETRY: WALL LENGTH

The third category of influencing parameters is related to the geometry. It entails both the dimensions of the wall and the presence of openings. The influence of openings has not yet been structurally investigated. These are burning topical issues for the construction industry. The formulation of advice with regard to movement joint schemes for facades with arbitrary opening patterns has to be properly planned. The cover of this report illustrates that numerical methods offer good prospects for the long term.

In this section we focus on the most important parameter of all: the wall length. The relationship between the crackfree wall length and imposed shrinkage is of immediate practical importance for the advice given with regard to movement joints. It is more correct to regard the wall length/height ratio as a variable [44], but the wall height is held constant ($H = 2.4$ m). The wall length L has been varied. The reference calculation from Section 6.3 concerned a length of 6 m. Additional calculations have been carried out for wall lengths of 4, 5, 8, 12 and 16 m. Figure 6.15 shows the shrinkage-crack width diagrams.

For relatively long walls (16, 12 and 8 m) it appears that almost immediately after crack initiation reflective cracking occurs. With the shortest wall (4 m) on the contrary, a large reserve is present before reflective cracking occurs. This difference is an immediate consequence of the stress distributions in the wall, as indicated in the

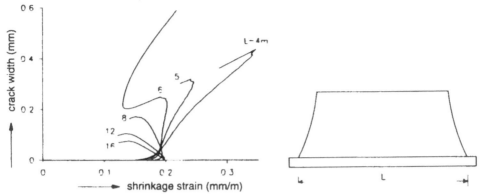

Figure 6.15. Influence of wall length on the shrinkage-crack width diagram. $R_a = 0.5$, $R_i = 1$, parameters as in Section 6.3.

Appendix. For long walls the stress distribution in the midsection is almost uniform, so that a crack will soon appear across the entire midsection. In short walls a gradient exists in the tensile stress distribution in the middle, which enables a redistribution with softening and gradual crack propagation to take place after crack initiation.

6.7 CONCLUSION: THE STEP TO CALCULATION RULES

Existing calculation rules give the crackfree wall length as a function of the imposed shrinkage (refer to Section 6.2.3). The crackfree wall length is interpreted as the wall length where for the given shrinkage, reflective cracking is just prevented from taking place. In the numerical calculations the wall length has been fixed and the shrinkage has been increased until at last reflective cracking occurs. This critical shrinkage can be measured in Figure 6.15 for different wall lengths. The relationship has been drawn in Figure 6.16 and leads to a 'numerically determined calculation rule'.

The numerical results fall within the range of the analytical data. The results show the greatest similarity with Copeland's formula [67] (Eq. 6.3), which was later on adopted by others. The vertical asymptote of this formula for an increasing wall length has been numerically confirmed. For an infinitely long wall the stress distribution in the midsection is uniform. By means of a simple elastic calculation it can be shown that the asymptote coincides with a shrinkage value $\varepsilon_k = f_t/(ER_a)$.

In the numerical result the transfer of the asymptote to the tougher behaviour of short walls is rather abrupt. The curve shows a rather sudden kink, while Copeland's formula shows a more gradual descent. To what extent this is realistic, depends on the realistic value of the chosen obstructions, adhesion properties in the joint between wall and foundation beam, etc. If, for example, the infinite bending stiffness of the obstruction $R_i = 1$ is replaced by $R_i = 0.06$, then the toughness of short walls is overestimated when compared to Copeland's formula, refer to Figure 6.17 [41]. Copeland's formula does not take into account the influence of bending stiffness and the properties of the joint between wall and foundation.

It is concluded that a link has been established between numerical simulations and

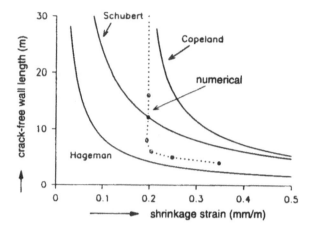

Figure 6.16. Crack-free wall length as function of imposed shrinkage. $R_a = 0.5$, $R_i = 1$, parameters as in Section 6.3. Comparison of numerical calculations from 6.6 with analytical formulae (Eqs 6.3-6.6).

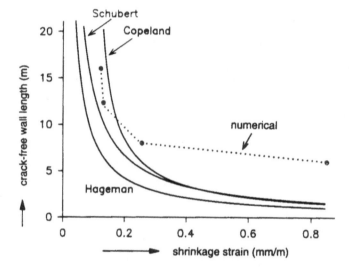

Figure 6.17. Crack-free wall length as function of imposed shrinkage. $R_a = 0.9$, $R_i = 0.06$, other parameters as in Section 6.3. Comparison of numerical calculations from [41] with analytical formulae (Eqs 6.3-6.5).

practical calculation rules. That was the main purpose of this project.

Because of the enormous amount and possible combinations of influencing parameters*, it is not yet possible to express an unequivocal judgement of the desirable and necessary adaptations of the formulae. The variation studies in the previous sections, however, are an important step in that direction. The understanding of the role of the various influencing factors has increased. Especially the influences found with regard to the bending-obstruction degree, axial degree of obstruction and friction/slip

---

* Suppose: There are 10 relevant influencing parameters, all of which can have three different values (a minimum, an average and a maximum). A complete investigation of all of them would at least require $3^{10} = 59049$ calculations (or experiments). In this first study a limited number of combinations have been investigated, usually for a wall length of 6 m. In subsequent research an extension of activities will take place in order to determine the correlation between the most critical influencing parameters.

along the joint between wall and foundation beam have to be translated into manageable factors in the formulae. Additional utilisation research has already been set in motion.

6.8 PRACTICAL RELEVANCE

In movement joint spacing calculations constructors use a certain design value for the shrinkage, dependent on the expected thermal and hygroscopical conditions of the wall. Thermal shrinkage is dependent on a number of factors such as the fact whether it concerns an outer or inner wall (this influences of course the expected difference in temperature), and with outer walls, the position of the wall in relation to the sun and the colour of the wall. Hygroscopical shrinkage will depend on the type of unit and mortar. For calcium silicate unit usually a higher value should be taken into account than for clay unit. Realistic calculation examples for south facades (outer walls) have among others been given by Klaas [72]. They include the following values:

– *Clay unit*
Thermal shrinkage 0.24 mm/m
(thermal expansion coefficient $\alpha = 6 \times 10^{-6}$ and difference in temperature 40°C)
Hygroscopical shrinkage 0.1 mm/m
Total: 0.34 mm/m
– *Calcium silicate unit*
Thermal shrinkage 0.24 mm/m (analogy clay unit)
Hygroscopical shrinkage 0.2 mm/m
Total: 0.44 mm/m

From the graphs as illustrated in Figure 6.16 and 6.17 the accompanying value of the crackfree wall length can be measured. This distance is equal to the required movement joint spacing distance. The term 'crackfree' in this consideration is linked to reflective cracking. If no reflective cracking occurs, than the wall is crackfree. In the analytical considerations such a definition has not explicitly been mentioned. Here we touch on the point that not only reflective cracking, but also a check on the crack width of the still incomplete crack is of practical importance. These data result from numerical simulations. For the reference wall from Section 6.3 the maximum occurring crack width increased from 0.24 mm before reflective cracking (bottom of wall) up to 0.55 mm immediately after reflective cracking (top of wall), refer to Figure 6.4. These values should be assessed on the basis of aesthetic and structural crack width requirements. Besides, plastered walls will be subject to more stringent criteria than non-plastered walls. A hidden crack in vertical or longitudinal joints is obviously less eye-catching than a crack in plasterwork.As for the advice given with regard to movement joints, the complex facades with random opening patterns often form bottlenecks. For repeatable massive walls it is possible to use standard formulae, graphs and tables. For unique walls with openings there is a need for simple software which includes the graphical wall geometry and the possibility to click on restraint values and material properties. On the basis of background research as described in this chapter, a movement joint scheme can then automatically be generated.

CHAPTER 7

Case study pier-main wall connections

7.1 INTRODUCTION

Stabilising facilities in stacked masonry structures consist of a cooperative system of main walls, piers and floors. In many cases the connection between pier and main wall is an important link. In traditional construction techniques this link is the result of unit laying in bond around the corner. In the modern stacking technique it is more practical and cheaper from the constructional and technical point of view to link the pier by means of a vertical glued joint to the main wall. The question is, however, whether such a connection meets the current requirements with regard to strength. An experimental investigation has given a decisive answer about that (Raijmakers & Van der Pluijm [57]).

This chapter deals with this matter from a numerical point of view. The approach is as follows. Firstly, the available experimental data are used to verify the numerical models. Numerical predictions of failure loads and failure mechanisms for glued pier-main wall connections are compared with the available experimental results. If the similarity is satisfactory, then the test series can subsequently be extended with 'numerical tests' in which geometry, boundary conditions and material parameters are varied. In this way the role of the various influencing parameters on the failure behaviour can be quantified, in analogy with the approach in the previous chapter.

Because the previous case study was extensively discussed, only a short description of the main principles of this case study will suffice. One example is discussed from the variation studies as a demonstration of the way in which calculation rules can be derived from the investigation. For extensive information refer to [38].

7.2 PRESENT-DAY KNOWLEDGE IN THE ANALYTICAL FIELD

From the limited quantity of literature it appears that there is little experience with manual calculation of failure conditions for pier-main wall connections. The Dutch Practical Guidelines 6791 talk of the transfer of a certain shear force per running meter height of the pier. This suggests that present-day calculation methods proceed from a constant shear stress value across the height. This is also mentioned in [57]. Such a method, however, is only a rough approach to reality. For, concentrated horizontal force initiation on top and bottom will lead to highly non-uniform shear stress values along the connection, with stress peaks on top and bottom. As will appear later on, this is confirmed by the numerical calculations. This deviation and the inherent complexity of the connection probably explains why manual calculation models are not or hardly available.

7.3 PRESENT-DAY KNOWLEDGE IN THE EXPERIMENTAL FIELD

This section gives a summary of the experimental research by Raijmakers & Van der Pluijm [57], of which the results have been released for usage and verification in this CUR project.

For the experiments five different types of test walls were built in calcium silicate unit. Each type of test was repeated in triplicate, so that in all fifteen connections have been tested. The construction parts are U-shaped, consisting of a main wall and two piers. The specimen varied according to unit size (blocks, elements and small-sized units) and type of connection (toothed versus non-toothed, non toothed ageing referred to as vertical line joint). In this chapter we focus on the three tests with element size calcium silicate units and a glued vertical line joint connection. The main wall has a thickness of 265 mm and the pier has a thickness of 100 mm. The height of the specimen is 2500 mm. The length of the piers is 600 mm. The observed length of the main wall is 1600 mm. This length corresponds with an assumed effective width of the main wall which has a total length of 9 m as used for standard terraced housing. Figure 7.1 shows the set-up.

The U-shaped installation simulates a situation as occurs when housing construction takes place in series. Here, two different loads are relevant: a vertical force, that is dependent on the size of the nave of the terraced houses and a horizontal force, perpendicular to the main wall, due to wind load on a front facade. Both forces are lower on the first floor than on the ground floor. In this research the values for the

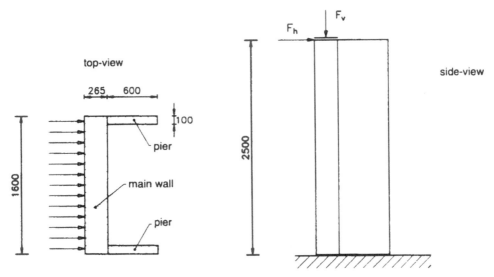

Figure 7.1. Geometry and loading at the tested connection of pier and the main wall.

ground floor were used, because they reflect the most critical case. The specimen were pre-loaded with a vertical force of 100 kN/m' main wall. Subsequently, a horizontal load F_h was applied. Both the vertical and the horizontal force are applied on top along the entire main wall. In addition, the dead weight of the construction is of importance. Especially with a view to verification of numerical models the tests have been carried out deformation-controlled. With this technique not only the failure behaviour but also the residual behaviour after failure could be registered. Three failure mechanisms can be distinguished (Fig. 7.2):

1. The construction as a whole tilts around the 'toe' of the pier.

2. The pier fails due to compression, with splitting (tensile) cracks in the compression diagonal.

3. The vertical line joint fails due to shear.

With a relatively low top load and/or very strong connections, Mechanism (1) occurs. The construction is still intact, but tilts as a whole which results in an unlimited increase of deformation under constant load (tilting equilibrium). In the toothed con-

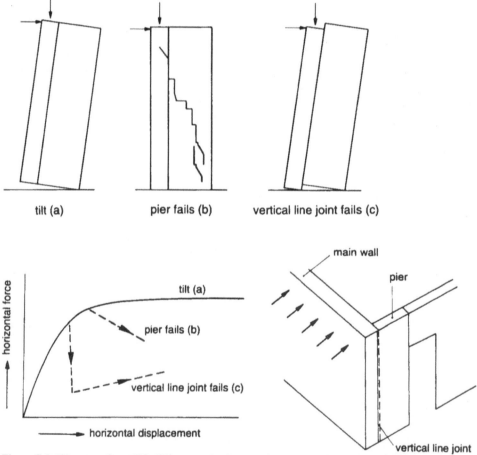

Figure 7.2. Diagram of possible failure mechanisms and accompanying relationship between horizontal load and horizontal displacement on top of the main wall. a) Composed pier-main wall construction tilts, b) Pier fails, c) Vertical line joint fails.

nections where the units have been laid in bond around the corner, cracks gradually appear along the compression diagonal in the piers. In some cases, the pier is able to withstand this and the tilting balance is then still reached. In other cases this leads to a limited decrease of the load that can be absorbed. The construction then fails according to Mechanism (2). The connections with vertical line joint usually failed at an earlier stage, due to shear of the vertical line joint, Mechanism (3). This was a sudden, brittle fracture which was accompanied by a quick reduction of the load that can be taken. As residual behaviour it was found that pier and main wall were standing adjacent to each other with a sheared-off joint and could still transfer a force through dry friction, be it with a large deformation. Two photographs of experiments that failed in respectively the vertical line joint and the pier are shown in Figure 7.3. The failure load in Mechanism (3) was usually lower than in Mechanism (2), which means that the perpendicular joint connection is less strong than the toothed connection. The vertical line joint connection, however, is so strong that the NPR requirement was met, which has been recorded in the TNO-Building Performance Test (TNO-Bouw Prestatietoets) [73].

7.4 NUMERICAL MODELLING

We will consider the glued pier-main wall connection with glued vertical line joint with geometry and loads from Figure 7.1. From preliminary calculations [38] it appeared that two-dimensional modelling with plane stress elements was accurate enough. The behaviour of main wall and pier is assumed linear-elastic. All the non-linear behaviour and crack formation was considered to be concentrated in the interface elements at the vertical line joint and the base joint. This modelling is accurate enough to simulate failure Mechanism (3) with shear in the vertical line joint, as observed here. The assumption of linear-elastic behaviour implies that possible local crack formation or crushing in the pier are not modelled. This has no consequences with regard to the predicted failure load, but could lead to a limited underestimation of the deformations at failure. This turned out better than expected.

In the base joint under the main wall and pier it was assumed that only contact pressure could be transferred, but no tension. In the experiments foil was applied between the (greasy) floor and pier-main wall. The assumption of no-tension behaviour is, therefore, realistic. For the vertical line joint, the Coulomb friction model with cohesion-softening from Figure 3.9 was applied. In the preliminary analyses the parabolic yield criterion from Figure 3.12a was tested as well, but this appeared to lead to an overestimation of strength and post-failure behaviour [38].

Table 7.1 shows the material parameters. The values were chosen on the basis of the companion experiments [57] that were carried out with the test series, the general micro-experiments from Chapter 2, and other experimental data [56]. The parameters are dependent on many factors and the three sources mentioned, for that reason, show differences. The set of parameters that was selected should be regarded as a compromise, in which attention was primarily paid to the first source because this one includes the data of the experiment in question. The stiffness values of the joints have been determined, departing from the real thickness (2 mm for base joint and 3 mm for perpendicular joint) and an elastic modulus of 1000 N/mm^2.

a

b

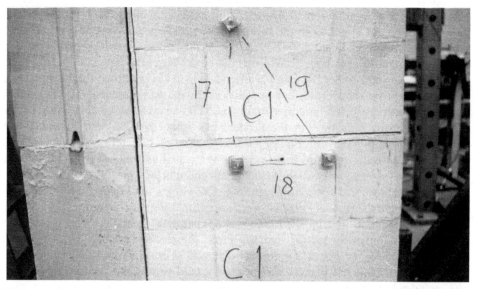

Figure 7.3. Experimental observations [57]. a) Toothed connection that fails in pier, mechanism (2) in Figure 7.2, b) Vertical line joint connection that fails in vertical line joint, mechanism (3) in Figure 7.2.

Table 7.1. Material parameters for pier-main wall connection with glued vertical joint.

Component	Parameter	Symbol	Value	Dimension
Calcium silicate	Elastic modulus	E	5000	N/mm^2
	Poisson's ratio	ν	0.12	–
	Unit mass	ρ	1800	kg/m^3
Base joint	Normal stiffness	k_n	10^6	N/mm^2
	Shear stiffness 'no-tension' behaviour	k_t	10^6	N/mm^2
Vertical joint	Normal stiffness	k_n	333	N/mm^2
	Shear stiffness	k_t	139	N/mm^2
	Cohesion	c_u	0.4	N/mm^2
	A Mode-II fracture energy	G_f^{II}	0.05	J/m^2
	Angle of friction	$\tan \phi$	0.75	–
	Angle of dilatancy	$\tan \psi$	0.1	–

In the calculation firstly the dead weight and the vertical top load were applied. Subsequently, the horizontal load was increased by small increments up to and including failure.

7.5 NUMERICAL RESULTS

The load-displacement diagrams are drawn in Figure 7.4. The load is the horizontal force on top of the main wall and the displacement is the corresponding horizontal displacement at that location. In the beginning, the experiments show a linear ascending part and subsequently a gradual decline of the curve to the maximum load. The maximum load in the experiments amounts to 42 kN on average. After the peak, a sudden drop of the load occurs in the three experiments, down to a residual level of about 14 kN. The failure load corresponds with the sudden appearance of a shear crack in the vertical line joint. The sharp drop is an indication of the fact that the test could not be held stable. The formation of the shear crack was accompanied by a bang and appeared to be a dynamic phenomenon. After the drop off a gradual rise can be observed in the diagrams, which, however, is far less steep than the rise in the start-up phase. This residual behaviour corresponds with separated tilting of main wall and pier which are standing against each other in dry friction, refer to Figure 7.2c.

The similarity between the numerical result and the experimental curves is very good. The numerical analyses also shows a linear ascending part of a curve that gradually declines. The chosen elastic modulus appears to be realistic because the initial gradient is similar to the experimentally determined gradient. The peak load amounts to 44 kN. After the peak both the load and the displacement appear to decrease. It now starts to be a bit boring for the reader: here too, a snap-back is involved once again. This phenomenon can occur because it concerns a quasi-static simulation, in which an arc-length method with the shear along the vertical line joint is applied as control parameter to keep the solution process stable (refer to Section 3.2.2). For an explanation of this type of behaviour refer to earlier examples in

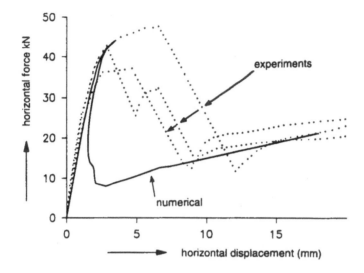

Figure 7.4. Comparison of experimental and numerical loaddis-placement diagrams.

this report. In the experiments explosive crack propagation occurred. It was not possible to keep the test stable and it was impossible to register the crack propagation and redistribution in the intermediate section between peak and residual level; the load suddenly dropped. Here, the numerical method gives additional information.

Also the residual behaviour appears to be surprisingly similar to the experiments. The residual stiffness represents a situation where the shear crack has fully developed, that is to say the Coulomb friction envelope with cohesion $c = c_u$ has reached its residual level when cohesion $c = 0$ due to softening. Main wall and pier are then standing next to each other without cohesion, and due to dry friction a force can still be transferred analogous to the experimental results. Both main wall and pier tilt around their toe. Also subtleties, among which the fact that the main wall was first completely lifted and came down again after the shear crack had originated to rest with its point on the ground, appear to be captured accurately by the numerical simulation.

Figure 7.5 and 7.6 show the incremental deformations and total deformations during various stages of the loading process. Incremental deformations are the deformations in one single load increment, so that the behaviour of the crack at that moment is clearly visible. Total deformations are the accumulated deformations from the beginning up to and including the last load increment that was carried out. The graphs reveal the initial behaviour, the propagation cracking process and the residual behaviour.

The crack is a pure Mode-II crack due to shear. In the literature on concrete it has been argued for several years whether Mode-II crack formation is physically possible or not. The calculation illustrates that with masonry, where a potentially weak plane is present in the form of a joint, the Mode-II mechanism can indeed occur. The cohesion c_u and its softening is the major material parameter in this process [38].

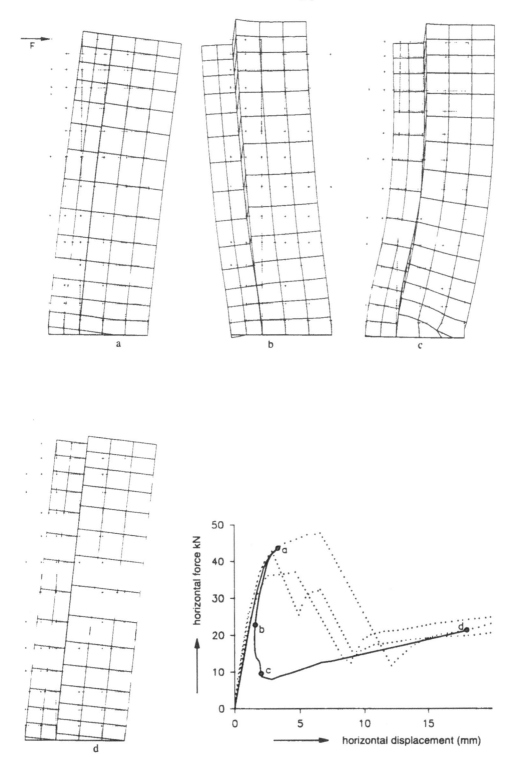

Figure 7.5. Incremental deformations at increasing stages of the loading process.

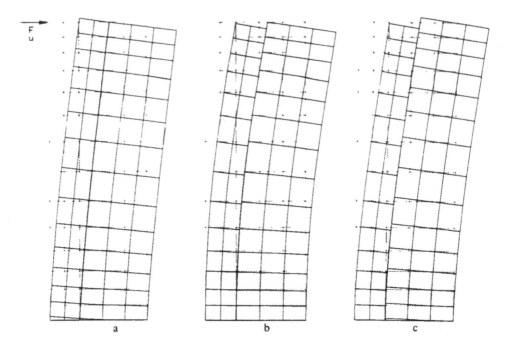

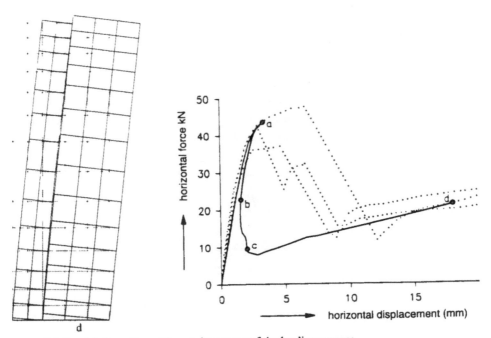

Figure 7.6. Total deformations at increasing stages of the loading process.

7.6 CONCLUSION: THE STEP TO CALCULATION RULES

Given the positive outcome of the verification, subsequently boundary conditions, geometry and material properties of the structure were varied in order to contribute to the development of rational calculation rules [38, 74]. An example is here dealt with in brief. This relates to the length of the pier. This was varied from 60 cm for the reference calculation, to 45 cm and 90 cm, which correspond with practical sizes which match dimensions of blocks or elements made out of calcium silicate units. In these calculations $c_u = 0.3$ N/mm^2 was used. Figure 7.7 shows the influence. The maximum load decreases and the brittleness of the structure increases with a decreasing pier length. The shear crack will develop faster. The relation between the failure load and the pier length has been shown diagrammatically in Figure 7.8. This can be interpreted as an initial impetus to a calculation rule. An engineer can use such diagrams to determine the required pier length for the design value of the horizontal wind load in question.

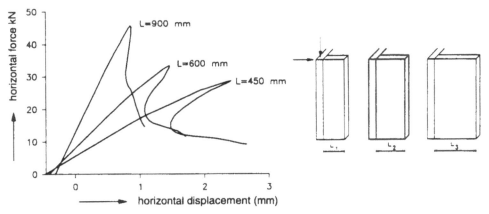

Figure 7.7. Influence of pier length on load-displacement diagrams.

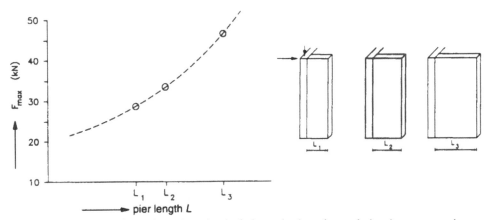

Figure 7.8. Example of practice-oriented calculation rule that gives relation between maximum load and pier length.

Just like in the previous case study, the accent was on the verification and scanning of numerical possibilities. In subsequent projects an extension will take place and practical calculation rules will be worked out for stability. As for the influencing parameters, especially the boundary conditions on top of the pier are relevant [74]. So far, the top edge was held completely free, but in reality floors made out of hollow-core slabs or solid slab floors are present which result in a counteracting moment. Other influencing factors are the rotational stiffness of the foundation, the three-dimensional effect of the building as a whole and its influence on the assumed effective width.

CHAPTER 8

Concluding remarks

An experimental/numerical basis has been created for the rational foundation of structural design in masonry. The results of material tests on tension, compression and shear (micro) are translated into numerical models, with which subsequently the non-linear structural behaviour (macro) can be simulated. The specific technical conclusions have been given in each chapter.

The research has led to a framework in which softening, friction, slip, dilatancy and snap-back behaviour are the dominant concepts for both tension and shear cracks. It has been shown that these concepts are not just academic issues, but a pressing necessity to come to a scientifically sound approach to crack formation, crack propagation and snap through reflective cracking in structures of unit like materials. The properties that are mentioned determine to a high degree the structural limits of the usually unreinforced masonry structures in the Netherlands.

The transformation of these research results to the construction practice takes place by means of the composition, improvement and foundation of design and calculation rules. With three case studies a step has been made in that direction. Simulation of the behaviour of walls under restrained shrinkage leads to practical guidelines on movement joint spacing to prevent or reduce crack formation. A numerical/experimental approach to pier-main wall systems adds to improved safety considerations and rational calculation methods for stability. The third case study concerned diaphragm walls and has been reported separately [6]. In subsequent projects the utilisation route will be further extended with the support of the unit-, block- and mortar industry, while progressive scientific research takes place with the support of the National Foundation of Technical Sciences (Stichting voor de Technische Wetenschappen, STW) among others.

Influence of wall length on stress distribution with restrained shrinkage

In the case study on crack behaviour of walls with restrained shrinkage the influence of the wall length on the tensile stress distribution in the midsection plays an essential role (refer to Sections 6.2.3 and 6.6). The wall length is of importance because it is linked to the issue of required movement joint spacings. The tensile stress distribution in the midsection is of importance with a view to crack formation, crack propagation and reflective cracking (snap-through).

This supplement gives a graphical impression of the linear-elastic tensile stress distributions with varying wall lengths. A wall height $H = 2.4$ m was departed from. The principal tensile stresses in the wall and the tensile stress distribution in the midsection A-A are shown, for wall lengths L of respectively:

16 m ($L/H = 6.67$),

12 m ($L/H = 5$),

8 m ($L/H = 3.33$),

6 m ($L/H = 2.5$),

4 m ($L/H = 1.67$).

Both the bending stiffness and the axial stiffness of the restraint are assumed to be infinite ($R_i = 1$ and $R_a = 1$). Linear-elastic behaviour is assumed with elastic modulus $E = 5000$ N/mm^2 and transverse contraction coefficient $v = 0.2$. On the walls a shrinkage ε_k of 0.0002 mm/m was imposed. With total restraint this would lead to a tensile stress equal to $E\varepsilon_k = 1.0$ N/mm^2, occurring in the midsection of the infinitely long wall. The principal tensile stresses are drawn in grey tones, where black corresponds with the maximum stress and white with the minimum stress. The longest wall leads to a nearly uniform tensile stress distribution in the midsection. As the wall length decreases the gradient of the tensile stress distribution in the midsection increases. For the shortest wall a compressive stress was even found on top. The implications for crack formation are dealt with in Chapter 6. With long walls the crack will be present across the entire midsection immediately after initiation. A reserve is present when the wall length decreases. The crack will start at the bottom and will subsequently slowly propagate upwards, depending on the gradient that is present and the extent of softening.

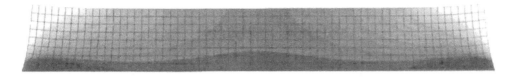

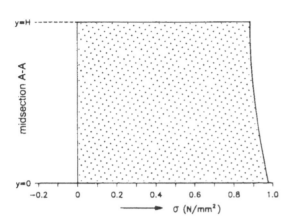

Wall length 16 meter

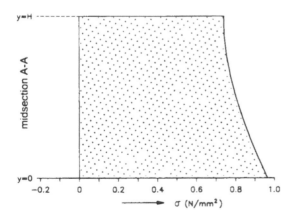

Wall length 12 meter

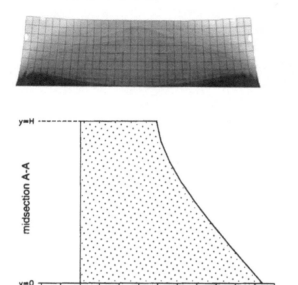

Wall length 8 meter

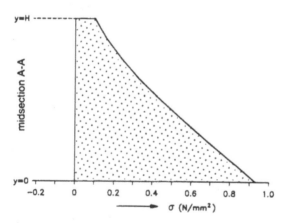

Wall length 6 meter

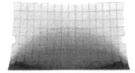

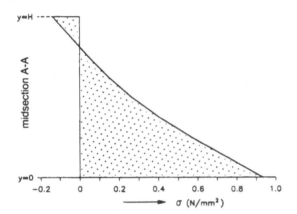

Wall length 4 meter

References

1. Siemes, A.J.M., Safety of Masonry Structures – Probabilistic approach. IBBC-TNO report no. B-85-588, Delft, 1985 (in Dutch).
2. CUR-report 87-3. Structural Masonry: Questions and Lack of Knowledge. Final report CUR pre-advisory committee PC 55. H.J.M. Janssen (ed.), CUR, Gouda, 1987 (in Dutch).
3. Schiebroek, C.J.M. (ed.). Final advice CUR programme-advisory committee, PAC 4, Structural Masonry (in Dutch).
4. CUR-report 90-6. Pre-advice Computational Masonry Mechanics, Final report CUR committee PA33 'Masonry Mechanics', H.J.M. Janssen, J.G. Rots & J.C. Walraven (eds), CUR, Gouda, 1990.
5. EUROCODE no. 6, Common unified rules for masonry structures. Draft report EUR 9888 EN, Commission of the European Communities, 1988.
6. CUR-report 94-2. Diaphragm walls in masonry. D.A. Hordijk & J.G. Rots (eds), CUR, Gouda, 1994 (in Dutch).
7. Van der Pluijm, R., Evaluation of Bond Tests on Masonry. TNO-Bouw report no. 93-CON-R1278, Delft, 1993.
8. Van der Pluijm, R., Shear Tests on Masonry, Inventarisation and Proposal. TNO-Bouw report no. BI-90-215, Delft, 1990 (in Dutch).
9. Vermeltfoort, A.Th. & Van der Pluijm, R., Deformation Controlled Tensile and Compression Experiments on Brick, Mortar and Masonry. TNO-Bouw/TU Eindhoven report no. B-91-0561, 1991 (in Dutch).
10. Vermeltfoort, A.Th., Wide versus Slender. Comparison of the Mechanical Properties in Compression of Wide and Slender Brick and Calcium-silicate Specimen. TU Eindhoven report no. TUE/BKO/92.04, 1992 (in Dutch).
11. Van der Pluijm, R., Deformation Controlled Micro-shear Tests on Masonry, TNO-Bouw report no. BI-92-104, Delft, 1992 (in Dutch).
12. Van der Pluijm, R., Uni-axial Deformation Controlled Tensile Tests on Bricks, Mortar and Masonry – Test results. TNO-Bouw report no. B-91-0560, Delft, 1991 (in Dutch).
13. Hordijk, D.A., Local Approach to Fatigue of Concrete. Dissertation, TU Delft, Faculty of Civil Engineering, 1991.
14. Groot, C.J.W.P., First Minutes Water Transport from Mortar to Brick. *Proc. 9th Int. Brick/Block Masonry Conference, Berlin*, pp. 71-78, 1991.
15. SBR-report 129, Properties of Masonry, Stichting Bouwresearch, Rotterdam, 1985 (in Dutch).
16. SBR-report B 19-2, Prefab Masonry Mortars. Stichting Bouwresearch, Rotterdam, 1979 (in Dutch).
17. Salin, S., *Structural Masonry*, Prentice-Hall, Englewood Cliffs, New Jersey, 1971.
18. Van der Haar, P.W. & Leeuwen, J., Literature Survey on Strength and Stiffness Properties of Masonry. IBBC-TNO report no. BI-78-44/62.2110, Delft, 1978 (in Dutch).
19. Riddington, J.R., Gambo, A.H. & Edgell, G.J., An Assessment of the Influence of Unit Aspect Ratio on Bond Shear Strength Values Given by the Proposed CEN Triplet Test. *Proc. 9th Int. Brick/Block Masonry Conference, Berlin*, pp. 1321-1328, 1991.

20. Stockl, S., Hofmann, P. & Mainz, J., A comparative Finite Element Evaluation of Mortar Joint Shear Tests. *Masonry International,* 3(3): 101-104.

21. Naninck, C. & Vermeltfoort, A.Th., The Development of Strength in Masonry Mortars. TU Eindhoven report no. TUE/BKO 92.03, 1992 (in Dutch).

22. NBN 24-301. Design and Calculation of Masonry. Belgisch Instituut voor Normalisatie (BIN), p. 10, 1980 (in Dutch).

23. Design NEN-EN 1052-1 (draft version prEn GGGG-1/1991 E), Test methods for masonry. Part 1: Determination of the compressive strength. Publication for comments only, Nederlands Normalisatie-Instituut (NNI), Delft, June 1993 (in Dutch).

24. Design NEN-EN 1052-3 (draft version prEn GGGG-3/1991 E), Test methods for masonry. Part 3: Determination of the initial shear strength. Publication for comments only, Nederlands Normalisatie-Instituut (NNI), Delft, June 1993 (in Dutch).

25. Beranek, W.J., Manual for the Mechanical Behaviour of Masonry. Part 1: Theory and Modelling. Report CUR-committee C77, Analytical Masonry Mechanics, Gouda/Delft, 1992 (in Dutch).

26. Design NEN-EN 772-1, Design Methods for Masonry Units. Part 1: Determination of the compressive strength. Publication for comments only, Nederlands Normalisatie-Instituut (NNI), Delft, July 1992 (in Dutch).

27. Rots, J.G., Computational Modeling of concrete fracture. Dissertation, TU Delft, Faculty of Civil Engineering, 132 pp., 1988.

28. De Borst, R., Computational methods in non-linear solid mechanics. Part 1: Geometrical non-linearity and solution techniques, Part 2: Physical non-linearity, Lecture notes, TU Delft report no. 25-2-91-2-06. TNO-Bouw report no. BI-91-043, TU Delft, Faculty of Civil Technique, Delft, 1991.

29. Ali, Sk.S. & Page, A.W., Finite Element Model for Masonry Subjected to Concentrated Loads. *ASCE Journal of Structural Engineering* 114(8): 1761-1784, 1988.

30. Beranek, W.J. & Hobbelman, G.J., Mechanics Models for Brick-like Materials. Part 1: Model for Material Behaviour, Part 2: Model for Structural Behaviour, Research reports CUR Committee C77, Analytical Masonry Mechanics, TU Delft, Faculty of Architecture, Delft, 1991 (in Dutch).

31. Eikman, S.M., Numerical Research on Cracking in Masonry. Part 1: Physically non-linear analysis of masonry, Part 2: Brick-mortar modelling on micro level. TU Delft report no. 25.2.91-2.19, TNO-Bouw report no. BI-91-206, Delft, 1991 (in Dutch).

32. Rots, J.G., Numerical simulation of cracking in structural masonry. *HERON* 36(2): 49-63, 1992.

33. Rots, J.G., Computer simulation of masonry fracture: Continuum and Discontinuum Models. *Computer Methods in Structural Masonry,* J. Middleton & G.N. Pande (eds), Books & Journals International, Swansea, UK, 93-103, 1991.

34. Coenraads, A.C., Detailed numerical analysis of masonry piers – Research on modelling aspects with DIANA. TU Delft report no. 25-2-91-2-07, TNO-Bouw report no. BI-91-073, Delft, 1991 (in Dutch).

35. Hillerborg, A., Modeer, M. & Petersson, P.E., Analysis of crack formation and crack growth in concrete by means of fracture mechanics and finite elements. *Cement and Concrete Research,* 6(6): 773-782, 1976.

36. Rots, J.G. & Lourenco, P.B., Fracture simulations of masonry using non-linear interface elements. *Proc. Sixth North American Masonry Conference,* A.A. Hahmid & H.G. Harris (eds), 2: 983-993, Drexel University, Philadelphia, USA, 1993.

37. Lourenco, P.B., Rots, J.G. & Blaauwendraad, J., Implementation of an interface cap model for the analysis of masonry structures. *Proc. EURO-C Conference on Computational Modelling of Concrete Structures,* N. Bicanic, R. de Borst & H. Mang (eds), Pineridge Press, UK, 1: 123-134, 1994.

38. Rots, J.G., Numerical research on pier-wall connections in masonry. TNO-Bouw report no. 93-CON-R0689, Delft, 1993 (in Dutch).

39. Lourenco, P.B. & Rots, J.G., On the use of micro-models for the analysis of masonry shear walls, *Proc. Second Int. Symp. on Computer Methods in Structural Masonry*, G.N. Pande & J. Middleton (eds), Swansea, UK, 1993.
40. Mann, W. & Muller, H., Failure of shear-stressed masonry: an enlarged theory, tests and application to shear walls, *Proc. British Ceramic Society*, 30: 223-235, 1982.
41. Rots, J.G., Numerical research on cracking in walls under restrained shrinkage. TNO-Bouw report no. 93-CON-R0688, Delft, 1993 (in Dutch).
42. Nauta, P., Substructuring, DIANA 5.1 User's Manual, Vol. 8, with example in DIANA Test Suite, TNO-Bouw, Delft, 1993.
43. NEN 6790, Brick Structures TGB 1990, basic requirements and methods of determination. Nederlands Normalisatie-Instituut (NNI), Delft, 1991 (in Dutch).
44. De Borst, R. & Nauta, P., Non-orthogonal cracks in a smeared finite element model. *Engineering Computations*, 2: 35-46, 1985.
45. Rots, J.G., Nauta, P., Kusters, G.M.A. & Blaauwendraad J., Smeared crack approach and fracture localisation in concrete. *HERON* 30(1): 1-48, 1985.
46. Rots, J.G., The smeared crack model for localized mode-I tensile fracture. *Numerical Models in Fracture Mechanics of Concrete*, F.H. Wittmann (ed.), Balkema, Rotterdam, pp. 101-114, 1993.
47. Feenstra, P.F., Computational aspects of biaxial stress in plain and reinforced concrete. Dissertation, TU Delft, Faculty of Civil Engineering, 151 pp., 1993.
48. De Borst, R., Simulation of strain localization: a reappraisal of the Cosserat Continuum. *Engineering Computations* 8: 317-331, 1991.
49. Rots, J.G. & Leonard, H., Long calcium-silicate walls – Brick/mortar and continuum modelling. CUR A33, Intermediate internal report, TNO-Bouw, 27 November 1991 (in Dutch).
50. Rots, J.G. & De Borst, R., Analysis of concrete fracture in 'direct' tension, *Int. Journal of solids & structures*, 25: 1381-1394, 1989.
51. Raijmakers, T.M.J. & Vermeltfoort, A.Th., Deformation controlled meso shear tests on masonry piers. TNO-Bouw/TU Eindhoven report no. B-92-1156, 1992 (in Dutch).
52. Vermeltfoort, A.Th. & Raijmakers, T.M.J., Deformation controlled meso shear tests on masonry piers. Part 2. Draft report, TU Eindhoven, dept. BKO, 1993.
53. Lourenco, P.B. & Rots, J.G., Understanding the behaviour of shear walls: a numerical review, *Proc. 10th Int. Brick/Block Masonry Conference*, N.G. Shrive & A. Huizer (eds), Masonry Council of Canada and the University of Calgary, 1: 11-20, 1994.
54. Schubert, P. & Weschke, K., Verformung und Risssicherheit von Mauerwerk, Mauerwerk-Kalender 1986, 145-159, Ernst & Sohn, Berlin, 1986 (in German).
55. Kasten, D. & Schubert, P., Verblendschalen aus Kalksandsteinen: Beanspruchung, rissfreie Wandlänge, Hinweise zur Ausführung. *Bautechnik* 3: 86-94, 1985 (in German).
56. Berkers, W.G.J. & Rademaker, P.D., Cracking in long walls – Final report, RCK report no. CK015.1, Research Centre Calcium Silicate Industry, Barneveld, 1992.
57. Raijmakers, T.M.J. & Van der Pluijm, R., Stability of calcium silicate piers, TNO-Bouw report BI-91-0219, Rijswijk, 1992 (in Dutch).
58. Erkens, S., Numerical investigation of cracking behaviour of masonry walls due to restrained shrinkage, TNO-Bouw report 94-CON-R0376, Delft, 1994 (in Dutch).
59. Cundall, P.A., A computer model for simulating progressive large scale movements in blocky rock systems. *Proc. Symp. of the International Society of Rock Mechanics, Nancy, France*, 1: II-8, 1971.
60. Janssen, H.J.M., Numerical mechanics of masonry. Subproject UDEC. Progress report 1990 CUR-committee A33, internal report TUE/BKO 91.05, 1991 (in Dutch).
61. Janssen, H.J.M., Numerical mechanics of masonry. Subproject UDEC. Progress report 1991 CUR-committee A33, report TUE/BKO 92.25, 1992 (in Dutch).
62. De Jong, P., Lessons from damage in building Industry (I), *Cement 1992*, 2: 26-28, 1992 (in Dutch).
63. Schubert, P., Zur rissfreien Länge von nichttragenden Mauerwerkswänden. Mauerwerk-Kalender 1988, 473-488, Ernst & Sohn, Berlin, 1988 (in German).

64. Ibrahim, K.S. & Suter, G.T., Contribution to the rational determination of movement joint spacing in concrete masonry walls. *Proc. Sixth North American Masonry Conference,* A.A. Hahmid & H.G. Harris (eds), 1: 491-503, Drexel University, Philadelphia, USA, 1993.

65. Bazant, Z.P. (ed.), *Fracture mechanics of concrete structures,* Part I: State-of-the-Art report on fracture mechanics of concrete-concepts, models and determination of material properties. Elsevier Applied Science, London and New York, 1992.

66. Rots, J.G., Computational case studies of size, shape and boundary effects. Proc. JCI Int. *Workshop on Size Effect in Concrete Structures,* H. Mihashi (ed.), Tohoku University, E & FN Spon, Chapman and Hall, London, pp. 335-350, 1994.

67. Copeland, R.E., Shrinkage and temperature stresses in masonry. *ACI Journal,* 53: 769-780, 1957.

68. Hageman, J.G., Study on shrinkage cracks. RCK report no. 189-1-0, 189-2-0, Barneveld, 1968 (in Dutch).

69. Schubert, P. & Glitza, H., Risssicherheit bei überwiegend horizontalen Formänderungen. Mauerwerk-Kalender 1983, 653-674, Ernst & Sohn, Berlin, 1983 (in German).

70. Schleeh, W., Die Zwängungsspannungen in einseitig festgehalten Wandscheiben. *Beton- und Stahlbetonbau,* 3: 64-72, 1962 (in German).

71. Kawamoto, T., Fundamental photo-elastic studies on shrinkage stresses in massive structures. *Proc. 8th Japanese National Congress for Applied Mechanics, Tokyo,* 1958.

72. Klaas, H., Bewegungsfugen in Verblendmauerwerk (Movement joints in facing masonry), *ZI* 2: 111-122, 1993.

73. TNO-Bouw, Performance test, Calcium silicate pier with vertical joint behaviour and strength properties, based on report no. BI-91-0219 [57], Rijswijk, 1992 (in Dutch).

74. Naaktgeboren, N.M. & Rots, J.G., Numerical research and variation studies on pier-wall connections of masonry, TNO-Bouw report no. 94-CON-R0881/1-2, 1994.

T - #0273 - 101024 - C0 - 254/178/9 [11] - CB - 9789054106807 - Gloss Lamination